**GCSE
SCIENCE
DOUBLE
AWARD**

CHEMISTRY

B Earl   L D R Wilford   Series editor K Foulds

D0709842

JOHN MURRAY

Other titles in this series:
GCSE Science Double Award Biology ISBN 0 7195 7157 X
GCSE Science Double Award Physics ISBN 0 7195 7159 6

International hazard warning symbols

You will need to be familiar with these symbols when undertaking practical experiments in the laboratory.

 Corrosive

 Oxidising

 Explosive

 Radioactive

 Harmful or irritant

 Toxic

 Highly flammable

 Eye protection

First published in 1996 by
John Murray (Publishers) Ltd
50 Albemarle Street
London W1X 4BD

Layouts by Eric Drewery.
Artwork by Barking Dog Art, Peter Bull Art Studio and David Farris.
Typeset in Rockwell Light and News Gothic by Wearset, Boldon, Tyne and Wear.
Printed and bound in Great Britain by Butler & Tanner Ltd, Frome and London.

A CIP record for this book is available from the British Library.

ISBN 0 7195 7158 8

**GCSE
SCIENCE
DOUBLE
AWARD**

CHEMISTRY

Contents

To the student   vii
Acknowledgements   viii

1 METALS

1.1 The elements – the building blocks   2
1.2 Metals in action   6
1.3 Chemical reactions   8
1.4 Metal olympics   12
1.5 Using the reactivity series   16
1.6 Extraction of metals from their ores   18
1.7 Iron   20

1.8 An expensive oxidation process   22
1.9 Rust prevention   24
1.10 Not what they seem   26
1.11 Aluminium   28
1.12 Copper   30
1.13 pH and all that!   32
1.14 Curing acidity   34
1.15 Salts   36
1.16 Bases   38
Exam questions   40

2 EARTH MATERIALS

2.1 Natural materials   44
2.2 Limestone – an essential mineral   46
2.3 The changing Earth   50
2.4 Journey to the centre of the Earth   54
2.5 Plate tectonics   58
2.6 Volcanoes   60
2.7 The atmosphere forms   62
2.8 New land from old   64
2.9 Metamorphic rocks   68
2.10 A cycle of change   70

2.11 Sorting rocks   72
2.12 Fossil fuels   74
2.13 Very useful   76
2.14 Fuels   78
2.15 The balance in the air   80
2.16 Only a cathedral   82
2.17 Alkanes   84
2.18 Hydrocarbon bonding   86
2.19 Giant molecules   88
Exam questions   90

3 STRUCTURE AND BONDING

3.1 Solids, liquids and gases 94
3.2 Changing state 96
3.3 Larger and smaller 98
3.4 The kinetic theory 100
3.5 Warming up and cooling down 102
3.6 The solution 104
3.7 Particles 106
3.8 What's in an atom? 108
3.9 Electron arrangement 110
3.10 Well held 112
3.11 X-raying substances 116
3.12 Covalent bonds 118
3.13 Giants 120
3.14 Metals 124
3.15 The chlor-alkali industry 126
3.16 Organising the elements 128
3.17 Electron structures and the periodic table 130
3.18 The alkali metals 132
3.19 Halogens 134
3.20 The periodic table – the remainder 136
Exam questions 138

4 PATTERNS OF CHEMICAL CHANGE

4.1 Chemical reactions 142
4.2 Helping industry 144
4.3 What happens when chemicals react? 146
4.4 What affects the rate of reaction? (1) 148
4.5 What affects the rate of reaction? (2) 150
4.6 Catalysts 152
4.7 Well brewed 154
4.8 Mouldy cheese 156
4.9 Cool reactions 158
4.10 Chemical energy 160
4.11 The smell of the country 164
4.12 Nitrogen 166
4.13 The need for nitrogen 168
4.14 Economy and production 170
4.15 Mass and moles 172
4.16 Moles, volumes and concentration 176
4.17 Calculating formulae 178
4.18 Moles and chemical equations 180
Exam questions 182

Glossary 185
Data tables 192
Index 196
Periodic table 200

To the student

This textbook has been written to help you in your study of chemistry as part of a GCSE science course. It includes information which covers the GCSE syllabuses of many examination boards and may provide more detail than you will need for your examination. So, you should check very carefully with your teacher whether there are any parts that are less relevant to your studies.

The book is divided into four sections which cover the major themes of chemistry. Within each section you will find that most topics are written at *foundation* level. You should read these topics fully. The remaining topics are written at a *higher* level, and you may not need to study all of these.

Each topic provides background information as well as explanations set in different contexts. This will allow you to think through the main ideas. The 'Things to do' sections draw on information provided within each of the topics. These sections will give you an opportunity to: answer the questions; show an understanding of the concepts; follow instructions that allow you to carry out practical work; and help you to investigate your own ideas.

This textbook provides a stimulus for you to find out more about chemistry. Any book can only give a certain amount of information and so you should try to develop confidence in searching for further information from a variety of sources both in and out of school. You should also use your curiosity to seek out further data to increase your understanding and to develop your investigations.

Examination questions are given at the end of each section so you can test your knowledge and understanding of the work covered. Higher level questions are denoted by a solid bar.

To help draw attention to the more important words, the first time scientific terms are used they are printed in **bold**. These words are included in a glossary at the back of the book. This glossary can be used as a self-test or as a simple reference section.

Try to read and re-read the relevant topics as often as you can so that you become familiar with the ideas and learn more effectively. Frequent revision will improve your understanding and recall of facts – and this will make further learning easier!

Finally, enjoy your science studies. Chemistry is a fascinating collection of facts and concepts which are underpinned by systematic scientific investigation. Chemistry is one aspect of science and one way of looking at ourselves in the world in which we live. We hope you enjoy your chemistry and using this book.

LDR Wilford and B Earl
St Aidan's Church of England High School, Harrogate

Acknowledgements

The authors would like to thank Irene, Katharine, Michael and Barbara for their never-ending patience and encouragement during the production of this textbook. In addition, thanks is given to Mr Dennis Richards, Headteacher, St. Aidan's Church of England High School, Harrogate, for his help and support. Also we wish to thank the editorial staff at John Murray for all their hard work and support.

The authors and publishers are grateful to Mr Vernon Hudson, Senior Teacher and Head of Science at Radcliffe High School, for his advice on the texts of *GCSE Science Double Award Biology, Chemistry* and *Physics*.

Exam questions are reproduced by kind permission of:
Midland Examining Group (MEG);
Northern Examinations and Assessment Board (NEAB);
Southern Examining Group (SEG);
University of Cambridge Local Examinations Syndicate (UCLES);
University of London Examinations and Assessment Council (ULEAC);
Welsh Joint Education Committee (WJEC).

The publishers have made every effort to trace copyright holders, but if they have inadvertently overlooked any they will be pleased to make the necessary arrangements at the earliest opportunity.

Photo credits

p.v *t* Ben Johnson/Science Photo Library, *b* Mike Hoggett/ICCE; **p.vi** *t* Crown Copyright. Historic Royal Palaces., *b* David George/Planet Earth Pictures; **p.2** NASA/Science Photo Library; **p.3** *t* U.S. Dept. of Energy/Science Photo Library, *b* Andrew Lambert; **p.4** *tl* Hackenberg-ZEFA, *tc* Andrew Lambert, *tr* ZEFA, *bl* Peter Sanders, *br* Ben Johnson/Science Photo Library; **p.5** Andrew Lambert; **p.6** *tl* Dr Jeremy Burgess/Science Photo Library, *tr* Richard Martin/Allsport, *b* ZEFA; **p.7** *l* ZEFA, *r* CEPHAS/TOP/Pierre Hussenot; **p.8** *t* CEPHAS/Mick Rock, *c* Nigel Cattlin/Holt Studios, *bl* John Mead/Science Photo Library, *br* Hanson Brick Limited; **p.9** *t* E.R. Degginger/OSF, *c* and *b* Andrew Lambert; **p.11** Andrew Lambert; **p.12** *tl* Andrew Lambert, *tr* Popperfoto, *b* Ancient Art and Architecture Collection; **p.13** *tl* Last Resort, *tr* Andrew Lambert, *c* ZEFA, *b* John Townson/Creation; **p.14** *t* Andrew Lambert, *c* and *bl* Andrew Lambert, *br* J. Pfaff-ZEFA; **p.16** Milepost 92 1/2; **p.17** Andrew Lambert; **p.18** *tl* Dr. B. Booth/Geoscience Features, *tr* Rex Features, *b* Ancient Art and Architecture Collection; **p.20** British Steel; **p.22** *tl* Chesnot/Rex Features, *tr* BLOK-ZEFA, *cl* Simon Fraser/Science Photo Library, *bl* Kairos, Latin Stock/Science Photo Library, *br* Ford Motor Company Limited; **p.24** *t* Ford Motor Company Limited, *c* Paul Brierley, *bl* John Townson/Creation, *br* Last Resort; **p.25** *t* Andrew Lambert, *b* Last Resort; **p.26** *l* John Townson/Creation, *r* Ronald R. Read; **p.28** *tl* Pascal Rondeau/Allsport, *tc* National Power plc, *tr* Last Resort, *b* ZEFA; **p.29** Anglesey Aluminium Metal Limited; **p.30** *t* ZEFA, *bl* UK Copper Plumbing and Heating Systems Board, *br* BICC Cables Limited; **p.31** Simon Fraser/Northumbria Circuits/Science Photo Library; **p.32** *t* and *c* John Townson/Creation, *b* Richard Packwood/OSF; **p.33** *t* John Townson/Creation, *b* Andrew Lambert; **p.34** *t* Biophoto Associates, *b* John Townson/Creation; **p.36** *t* ZEFA, *c* and *b* John Townson/Creation; **p.38** *t* and *b* Andrew Lambert; **p.39** Andrew Lambert; **p.43** Mike Hoggett/ICCE; **p.44** *t* and *cl* Ancient Art and Architecture Collection, *cr* photo: Mary Rose Trust, *bl* and *bc* Geoscience Features, *br* CEPHAS/TOP/Marc Tulane; **p.46** *t* Nigel Cattlin/Holt Studios, *b* Rex Features; **p.47** Robert Harding Picture Library; **p.48** *t* and *c* Andrew Lambert, *b* Tilcon Limited; **p.50** *t* Geoscience Features, *b* John Cleare/Mountain Camera; **p.51** *tl* and *b* Geoscience Features, *tr* Still Pictures; **p.52** *t* and *c* Geoscience Features; **p.53** *tl*, *tr*, *bl* and *br* Scottish Natural Heritage/L. Gill; **p.58** California Institute of Technology; **p.60** Lon Stickney/Rex Features; **p.61** *tl* Northern Ireland Tourist Bureau, *tr*, *cl*, *cr*, *bl* and *br* Geoscience Features; **p.62** NASA/Science Photo Library; **p.63** The Natural History Museum, London; **p.64** Geoscience Features; **p.65** Geoscience Features; **p.66** *t* ZEFA, *bl* Geoscience Features, *br* The Bridgeman Art Library; **p.68** Geoscience Features; **p.69** *t* David W. Jones/Lakeland Life Picture Library, *bl* and *br* Geoscience Features; **p.70** Tom Leach/OSF; **p.72** *tl* Kathie Atkinson/OSF, *tr*, *cl*, *cr*, *bl* and *br* Geoscience Features; **p.73** *tl*, *tc* and *tr* Andrew Lambert, *bl* and *br* reproduced courtesy of Volvic Natural Mineral Water; **p.74** John Townson/Creation; **p.76** Stevie Grand/Science Photo Library; **p.78** *t* Dr Mueller-ZEFA, *cl* British Gas plc, *cr* National Power, *b* Rex Features; **p.80** Popperfoto; **p.81** Planet Earth Pictures; **p.82** Mike Hoggett/ICCE; **p.83** Mike Reid; **p.84** *t* Calor Gas Limited, *cl*, *cr*, *bl* and *br* John Townson/Creation; **p.86** *t*, *bl*, *bc* and *br* Andrew Lambert; **p.87** *t*, *c* and *b* Andrew Lambert; **p.88** *t* and *b* Andrew Lambert, *c* John Townson/Creation; **p.93** Crown Copyright. Historic Royal Palaces.; **p.94** *t* Haydn Jones/Rex Features, *bl* Last Resort, *br* Rex Features; **p.96** *tl* A.N.T./NHPA, *tc* Brian Kenney/Planet Earth Pictures, *tr* Richard Coomber/Planet Earth Pictures, *c* Andrew Lambert, *b* David Hughes/Robert Harding Picture Library; **p.98** *t* United Press International, *b* Alex Bartel/Science Photo Library; **p.100** *tl* and *tr* Andrew Lambert, *bl*, *bc* and *br* John Townson/Creation; **p.102** *l* Last Resort, *r* Simon Fraser/Science Photo Library; **p.103** Martin Dohrn/Science Photo Library; **p.104** *t* Last Resort, *bl* SmithKline Beecham, *bc* Andrew Lambert, *br* Penhaligon's; **p.105** *tl* and *tr* Andrew Lambert, *b* Last Resort; **p.106** *t* Ancient Art and Architecture Collection, *b* Nils Jorgensen/Rex Features; **p.112** *t* Crown Copyright. Historic Royal Palaces, *b* Andrew Lambert; **p.113** *t* Gerard Vandystadt/Agence Vandystadt/Allsport, *cl*, *cr*, *bl* and *br* Andrew Lambert; **p.115** Topham Picture Point; **p.116** *l* Dept. of Clinical Radiology, Salisbury District Hospital/Science Photo Library, *r* Science Source/Science Photo Library; **p.118** *t* and *c* Last Resort, *b* Andrew Lambert; **p.119** *t* and *c* Andrew Lambert, *b* Last Resort; **p.120** Last Resort; **p.121** *t* Du Pont (U.K.) Limited, *c*, *bl* and *br* John Townson/Creation; **p.122** *t* Gilles Levent/Rex Features, *tl* Last Resort, *tr* Carole Shayle/ICCE; **p.123** John Townson/Creation; **p.124** *l* David Parker/Science Photo Library, *r* Rex Features; **p.125** Popperfoto; **p.126** *l* and *c* John Townson/Creation, *r* Last Resort; **p.128** *t* HMV UK Limited, *b* Science Photo Library; **p.129** *tl* Andrew McClenaghan/Science Photo Library, *tr* Martin Dohrn/Science Photo Library, *cl* Shout Pictures, *bl* John Townson/Creation, *br* Manfred Kage/Science Photo Library; **p.131** John Townson/Creation; **p.132** *tl* Martin Bond/Science Photo Library, *tr* John Townson/Creation, *c* NASA/Science Photo Library, *b* Andrew Lambert; **p.134** *t* Robert Harding Picture Library, *cl* Lupe Cunha, *cr* Last Resort, *b* Andrew Lambert; **p.135** Last Resort; **p.136** *tl* Roger Ressmeyer, Starlight/Science Photo Library, *tr* Ivor Walton, *b* Rex Features; **p.137** Last Resort; **p.140** David George/Planet Earth Pictures; **p.142** *t* Milton Heiberg/Science Photo Library, *bl* Last Resort, *br* Wine Magazine/CEPHAS; **p.143** *tl* Last Resort, *tr* Ancient Art and Architecture Collection, *b* Alan Towse/Ecoscene; **p.144** Philip Harris Ltd; **p.145** Philip Harris Ltd; **p.146** *l* Geoscience Features, *r* Last Resort; **p.148** *tl* and *tr* Last Resort, *b* Andrew Lambert; **p.149** *t* and *b* Andrew Lambert; **p.150** *t* David George/Planet Earth Pictures, *bl* and *br* Last Resort; **p.151** Ancient Art and Architecture Collection; **p.152** *t* Scott Bader Company Limited, *b* Rex Features; **p.154** *t* Nigel Cattlin/Holt Studios, *c* Itona Products, *b* CEPHAS/Stuart Boreham; **p.155** CEPHAS/TOP/Pierre Hussenot; **p.156** *t* John Townson/Creation, *ct*, *cb* and *b* Last Resort; **p.158** *t* Northern Ireland Tourist Bureau, *c* and *b* Last Resort; **p.159** *l* Biomark Products, *r* David Rogers/Allsport; **p.160** ZEFA; **p.163** John Townson/Creation; **p.164** *t* Nigel Cattlin/Holt Studios, *b* Andrew Lambert; **p.165** *t* Philip Steele/ICCE, *b* Philip Harris Ltd; **p.166** Nigel Cattlin/Holt Studios; **p.167** Nigel Cattlin/Holt Studios; **p.168** Simon Fraser/Science Photo Library; **p.170** ICI Billingham; **p.171** *t* ICI Billingham, *bl* and *br* Andrew Lambert; **p.174** Andrew Lambert; **p.176** Camping Gaz (GB) Limited; **p.179** James Holmes/Oxford Centre for Molecular Sciences/Science Photo Library; **p.180** Tilcon Limited.

(*t* = top, *b* = bottom, *r* = right, *l* = left, *c* = centre)

1

METALS

The elements – the building blocks

The universe contains millions of substances. Each one is made up of particles called **atoms**. Atoms are much too small to be seen – about 20 000 000 of them laid end to end would just cover the length of your small finger nail. Even so, we now know that each atom is made up from many much smaller particles – **sub-atomic particles**.

Elements

Each of the substances on earth is made up of the atoms of simple substances called **elements** – substances which contain only one type of atom. Because they contain only one type of atom they cannot be broken down into simpler substances. Some naturally occurring metallic elements, such as gold, have been used for thousands of years.

We now know of 112 different elements of which only 92 occur naturally. They range from some very reactive gases such as fluorine and chlorine, to the unreactive elements platinum and mercury. Others have been made artificially and are **radioactive** (see *GCSE Science Double Award Physics*, Topic 4.15) to some degree. They include elements such as curium and plutonium.

Every element has its own chemical symbol – a form of shorthand

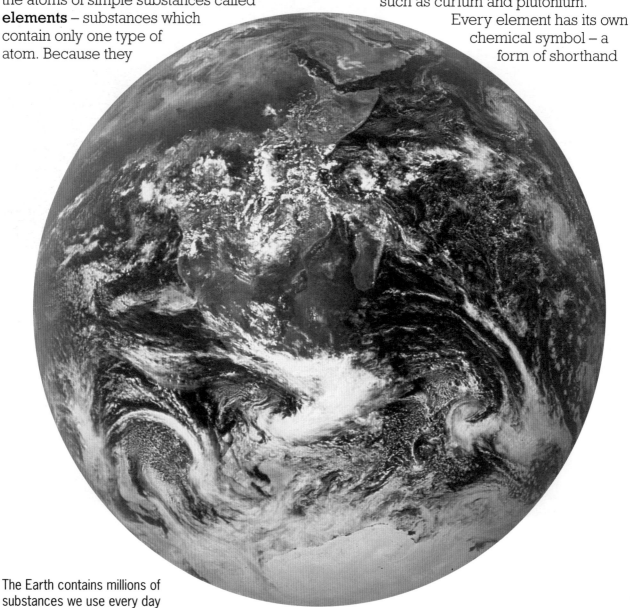

The Earth contains millions of substances we use every day

recognised throughout the world. Iron, for example, has the symbol Fe, whilst copper has the symbol Cu. The symbols for all the elements can be found in the **periodic table** on page 128.

Millions of other substances, called **compounds**, also exist. Compounds are substances which are formed when the atoms of several different elements combine in a particular way. Most naturally occurring substances are compounds. Many of them are the **raw materials** from which we extract elements which are not naturally occurring.

The names of compounds can also be written using the same symbols used to describe the names of elements. Copper sulphide, for example, contains copper and sulphur atoms. Its formula is CuS. The properties of elements are different to those of their compounds (see page 107).

Metals and non-metals

All the elements can be grouped according to their various properties. The simplest way to sort them is to classify them as **metals** and **non-metals**. Physical data of some metal and non-metal elements is shown in the table.

You will notice that generally metals have high densities, high melting points and high boiling points. Non-metals, on the other hand, have low densities and low melting and boiling points.

Plutonium is a radioactive element used in nuclear reactors

Some metals: (*from left to right*) aluminium, mercury and nickel

Physical data of some metal and non-metal elements

Element	Symbol	Metal/ non-metal	Density /gcm^{-3}	Melting point /°C	Boiling point /°C
Aluminium	Al	Metal	2.70	660	2580
Iron	Fe	Metal	7.87	1535	2750
Nickel	Ni	Metal	8.90	1453	2732
Copper	Cu	Metal	8.92	1083	2567
Hydrogen	H	Non-metal	0.07[a]	−259	−253
Sulphur	S	Non-metal	2.07	113	445
Nitrogen	N	Non-metal	0.88[b]	−210	−196
Chlorine	Cl	Non-metal	1.56[c]	101	−34

[a] at −254 °C
[b] at −197 °C
[c] at −34 °C

Source: Earl B, Wilford LDR, *Chemistry Data Book*, Nelson Blackie, 1991

A summary of the different properties of metals and non-metals is given in the table below.

The different properties of metals and non-metals

Property	Metal (e.g. copper)	Non-metal (e.g. sulphur)
Physical state at room temperature	Usually solid	Solid, liquid or gas
Malleability (how easily it can be shaped)	Good	No – usually soft or brittle when solid
Ductility (how easily they can be drawn into wires)	Good	No – usually soft or brittle when solid
Appearance (solids)	Shiny (especially when freshly cut)	Dull
Melting and boiling points	Usually high	Usually low
Density	Usually high	Usually low
Conductivity (thermal and electrical)	Good	Very poor

Metals are tough and strong

Chlorine is used in swimming pools, sticks of sulphur, and neon used in advertising signs

Malleable means that the metal can be formed into different shapes

Ductile means that the metal can be drawn into rods and wires using these shaper disks

★ THINGS TO DO

1 Make a list of all the elements you have seen or heard about. There may be some in your classroom which you can look at. For each element write its name, symbol and a description of its appearance and any other properties you can identify.

2 A group of pupils tested the properties of three unknown materials. They also carried out some research in their library. This is what they wrote about each one:

   Element A: Solid; yellow; breaks easily when hit with a hammer; smaller particles are crumbly. Melting point is 113 °C. Burns easily in air showing a blue flame and produces a dangerous gas. Density 1.96 kg/m³. Poor thermal conductor. Electricity does not pass through it.

   Element B: Dull grey colour. Soft and easily bent. When bent regularly it became warm and fractured. At the edges of the fracture it was bright and shiny. Melting point is 600 °C. Does not burn. Conducts electricity and heat passed along it fairly quickly. Density is 11.3 kg/m³.

   Element C: Dull brown colour. Bends fairly easily but not as easily as *B*. When cut open, the edges were bright and shiny orangey brown. Electricity passed through it and heat passed quickly through a rod made of the material. Did not burn. Its melting point is 1083 °C, and in its liquid state it boils at 2567 °C. Density = 8.9 kg/m³.

   Which of the elements were metals and which were non-metals? Explain your answer.

3 Complete the following paragraph about the metal elements choosing from the words given below.

   electricity, densities, liquid, high, lustrous, ductile, good, malleable, solids

   Metals usually have _____ melting and boiling points as well as _____. They are _____ conductors of heat and _____. They are usually _____ except in the case of mercury which is a _____. They are _____ as well as easy to draw into wires (_____).

4 The density of a material is found using the equation

$$\text{density} = \frac{\text{mass}}{\text{volume}}$$

Plan and carry out a series of tests to find the densities of different metals.

Metals in action

Metals have properties which make them perhaps the most versatile of all the materials used today. Their properties determine how they are used.

The correct choice of metal is particularly important in construction work. The metal chosen to build aircraft, for example, must be light and strong. Both aluminium and magnesium have these properties, but rather than use either one or the other, it is an **alloy** – a mixture of these metals called duralumin – which is used. Duralumin is much stronger than pure aluminium and is therefore much more suitable for aircraft construction.

The metal used in these cables must be able to conduct electricity and be fairly light and strong

The metal used in this bicycle frame should be light yet strong enough to withstand bumps

Alloying strengthens a metal by altering its structure by adding atoms of another element which are a different size. These act as a barrier and prevent the layers slipping over one another (see page 125).

Alloys have been designed to suit a wide variety of different needs. Many thousands of alloys are now made, the majority of which are 'tailor-made' to have properties which meet the needs of a particular job. The table shows some of the more common alloys, together with some of their uses.

The metal used in these aircraft must be light and strong

Alloys and their uses

Alloy	Composition	Properties	Uses
Bronze	90% copper, 10% tin	Harder than pure copper	Castings, machine parts
Cupro-nickel	30% copper, 70% nickel or 75% copper, 25% nickel	Stronger and harder wearing than pure copper	Turbine blades Coinage metal
Duralumin	95% aluminium, 4% copper, 1% magnesium, manganese, iron	Light but stronger than pure aluminium	Aircraft construction, and bicycle parts
Magnalium	70% aluminium, 30% magnesium	Stronger than pure aluminium	Aircraft construction
Pewter	30% lead, 70% tin, a small amount of antimony	Less malleable than pure lead	Plates, ornaments and drinking mugs
Solder	70% lead, 30% tin	Lower melting point than either element	Connecting electrical wiring

Steel, which is a mixture of iron (a metal) and carbon (a non-metal), is also considered to be an alloy. Of all the alloys we use, it is perhaps the most important. In fact, most of the iron which is made (see page 21) is changed into different steel alloys. The molten iron from the furnace is run into large containers called torpedoes. These are taken to a steel making plant where some of the carbon in the iron is removed and other metals such as chromium, nickel and manganese are added to form the different types of steel.

Different types of steel have quite different properties. The properties depend on the amounts of carbon and other metals which are present. The composition and uses of different types of steel are shown below.

Molten iron from the blast furnace being transferred to the steel making plant

In stainless steel the chromium prevents corrosion while the nickel makes it harder

Different types of steel

Steel	Typical composition	Properties	Uses
Mild	99.5% iron, 0.5% carbon	Easily worked, lost most of its brittleness	Car bodies, large structures
Hard	99% iron, 1% carbon	Tough and brittle	Cutting tools
Manganese	87% iron, 13% manganese	Tough, springy	Drill bits, springs
Stainless	74% iron, 18% chromium, 8% nickel	Tough, does not stain	Cutlery, surgical instruments, kitchen sinks

★ THINGS TO DO

1 Prior to 1992, 'copper' coins were an alloy consisting of 97% copper, 2.5% zinc and 0.5% tin. Use a magnet to compare 2p coins made before and after that date. What do you think is the difference between the alloys used? Try to find out the composition of the alloy used since 1992 and use the information to explain what you observed when you used the magnet.

2 'Many metals are more useful to us when mixed with some other elements.' Discuss this statement with respect to stainless steel.

3 Brass is an alloy formed by mixing copper and zinc.
a) Plan a series of tests which you could do to compare some of the properties of each metal – copper, zinc and brass – such as their strength, density, thermal and electrical conductivity, reactions with water and acids and so on. Carry out your tests under the supervision of your teacher.
b) Find out more information about each of the metals from the library. Draw up a table comparing the properties of each metal and try to explain any differences between the alloy and the elements from which it is made.

Chemical reactions

The photographs show a variety of chemical changes which you may have seen. Some are caused by the action of heat, others by living things (micro-organisms) and others by chemicals which react with one another.

They are described as **chemical changes** because chemicals inside the substances change permanently. When, for example, soft clay (which is of little use in its natural state) is strongly heated in a kiln, the clay changes to form bricks (which are much more useful to us).

$$\text{clay} \xrightarrow[\text{process}]{\text{manufacturing}} \text{bricks}$$
$$\text{(raw material)} \qquad\qquad \text{(useful product)}$$

Before the clay is fired it is soft and brown. After firing it is hard and red. During the process of 'firing' the clay, a new substance has been produced. Because the substances within the clay have been chemically changed it is not possible to reverse the process – we cannot change the bricks back into clay. They have *changed permanently*. A chemical reaction has occurred. The change in this case was brought about by the action of heat.

Heat changes dough into more edible bread

The decay on this fruit is caused by the action of micro-organisms

The iron is changed into rust as it reacts with water and oxygen in the atmosphere

Soft clay can be formed into blocks which when heated in a kiln are changed into bricks

Making use of chemical changes

Some of the chemical changes shown are beneficial, such as the one which changes dough into bread. Others are detrimental – decay spoils the fruit. If we know how chemical reactions work we can use that knowledge to make more useful (and sometimes more valuable) products from less useful raw materials and take steps to prevent the detrimental changes happening (such as when we refrigerate food to slow down decay).

From rock to metal!

Almost three-quarters of the elements are metals but only gold, silver and platinum are found in the earth as metals. Others must be extracted from compounds found in the Earth's crust.

Millions of years ago, liquids and hot gases from deep inside the Earth were forced into cracks in the Earth's crust. As the gases and liquids cooled they formed crystalline veins containing chemical compounds. Some contained metals such as gold. Rocks which contain important compounds which we need are called **minerals**. Some minerals are rich in metal compounds – they are called **ores**.

One of the ores found in the rocks of the Pennines is malachite. Malachite is a compound of copper – a chemical called copper carbonate – in which copper atoms are combined with carbon and oxygen atoms. This is the raw material from which, using a knowledge of chemical reactions, we can extract copper metal.

The copper carbonate is first changed into copper oxide by heating it in air. The resulting products are copper oxide and carbon dioxide, both of which are compounds. An easy way to describe the reaction is to use a word equation. Word equations contain the names of the **reactants** and the names of the **products**. An arrow is used to show that the reactants change into the products during the reaction.

The metal ore – malachite

Heat green copper carbonate and black copper oxide is produced

$$\text{copper carbonate} \xrightarrow{\text{heat}} \text{copper oxide} + \text{carbon dioxide}$$

This particular reaction is an example of **thermal decomposition**. Thermal decomposition occurs when one substance is heated (thermal) and splits up to form two or more simpler substances (decomposition). You will learn more about reactions which involve thermal decomposition in Topic 1.7, page 20.

The copper oxide can then be heated strongly with charcoal (a form of carbon). During the reaction the carbon atoms take the oxygen atoms from the copper oxide, forming carbon dioxide gas. Because the oxygen atoms have been removed from the copper oxide, only copper atoms are left. The word equation for this reaction is:

$$\text{copper oxide}_{(s)} + \text{carbon}_{(s)} \xrightarrow{\text{heat}} \text{copper}_{(s)} + \text{carbon dioxide}_{(g)}$$

(the reactants)   (the products)

The letters 's' and 'g' are in place of the words 'solid' and 'gas' respectively. These are the physical state symbols. If the reactant or product is a liquid then 'l' is used. If the reaction takes place in water (aqueous) solution then 'aq' is used.

In the topics which follow you will see how many different types of chemical reaction are used to make the materials we use every day. In particular you will learn that there are patterns which will help you predict the products of reactions.

An atomic view of a reaction

The particles of copper carbonate each contain one copper atom bound to one carbon atom and three oxygen atoms.

Because the symbol for copper is Cu, and the symbols for carbon and oxygen are C and O respectively, copper carbonate can be represented, in shorthand form, as $CuCO_3$. This is the chemical formula of copper carbonate, and shows the relative numbers of each atom in the compound.

	Cu	C	O$_3$
Ratio of atoms	1 :	1 :	3
Total number of atoms present = 5			

A particle view of the reactions described above shows some interesting features of all chemical reactions – that as some particles split up, others join together. During the first part of the reaction, the copper carbonate splits up and each copper atom joins with one oxygen atom. The carbon atoms recombine with two oxygen atoms forming carbon dioxide.

If we replace the chemical names of the reactants and products by their symbol or formulae we can produce a full **chemical equation**. The word equation for the thermal decomposition of copper carbonate can be replaced by:

$$CuCO_{3(s)} \xrightarrow{\text{heat}} CuO_{(s)} + CO_{2(g)}$$

Notice that the number of atoms on the left-hand side of the equation is the same as the number of atoms on the right-hand side. The same atoms are present in the products as were present in the reactants. This is a balanced chemical equation. Equations such as these tell us not only the names of the products and the reactants, but also the proportions in which they react.

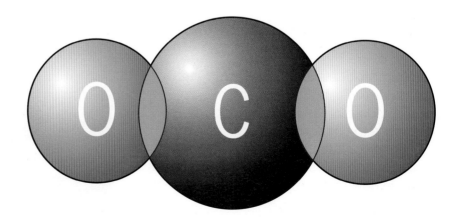

Carbon dioxide gas contains one carbon atom and two oxygen atoms

Magnesium reacts with oxygen to form the white powder, magnesium oxide

their physical state symbols. The atoms of some non-metal elements, such as oxygen, are joined together in pairs and form what we call **diatomic molecules** (see page 106). The formula for oxygen gas is therefore O_2.

$$Mg_{(s)} + O_{2(g)} \xrightarrow{\text{heat}} MgO_{(s)}$$

In the equation there are two oxygen atoms on the left-hand side (O_2) but only one on the right (MgO). We cannot change the formula of magnesium oxide, so to produce the necessary two oxygen atoms on the right-hand side we will need 2MgO – this means two particles of MgO are produced. The equation now becomes:

$$Mg_{(s)} + O_{2(g)} \xrightarrow{\text{heat}} 2MgO_{(s)}$$

To balance up the number of magnesium atoms on the left-hand side with those on the right-hand side we place a '2' in front of the magnesium. We now have a balanced chemical equation.

$$2Mg_{(s)} + O_{2(g)} \xrightarrow{\text{heat}} 2MgO_{(s)}$$

This balanced chemical equation shows us that two atoms of magnesium react with one molecule of oxygen gas when heated to produce two particles of magnesium oxide.

Let's try another example. The photograph shows magnesium metal burning brightly in air. During this reaction the magnesium reacts with oxygen gas. The white powder produced is called magnesium oxide. The word equation is:

$$\text{magnesium} + \text{oxygen} \xrightarrow{\text{heat}} \text{magnesium oxide}$$

We can replace the words with symbols for the reactants and the products and include

---

★ THINGS TO DO

1 Explain the meaning of the terms:
   a) chemical reaction;
   b) chemical change;
   c) thermal decomposition.

2 Which of the following involve chemical changes?
   a) melting ice cubes;
   b) burning a candle;
   c) adding sugar to a cup of coffee;
   d) expansion of telephone wires;
   e) baking cakes.

3 Talk with others in your group and add another five chemical reactions which you have seen.

For each one describe what happens. If you can, write word equations for each reaction.

4 Balance the following equations and include any missing state symbols:
   a) $Mg_{(s)} + CO_{2(g)} \rightarrow MgO + C$
   b) $Ca + H_2O \rightarrow Ca(OH)_{2(aq)} + H_2$
   c) $Na + Cl_2 \rightarrow NaCl$

5 Plan an experiment to find the mass of oxygen taken from the air when magnesium is burned. Carry out your experiment, under the supervision of your teacher, and write a brief description of what you did and what you found out.

Metal olympics

Although metals have similar physical properties they differ in the way in which they react with other substances. Consider, for example, what happens to three different metals when they are exposed to the air.

Potassium is a very soft metal which reacts violently (and immediately) with both air and water. It is stored under oil to prevent this type of reaction occurring. Iron also reacts with air and water, but much more slowly forming rust. Eventually all of the original iron will have changed to rust. Gold does not react with either water or air – it remains totally unchanged after many hundreds of years.

Potassium is said to be more reactive than iron and in turn iron is said to be more reactive than gold.

To help us understand more about the ways in which metals react (such as which substances they are likely to react with, and under what conditions) we need to know more about their reactivity.

Potassium burns violently

Iron rusts slowly

Gold does not change

Metal reactions

Before deciding on a particular metal for a job we must have an idea about its physical properties (such as those described on page 4) and whether, and how, it will react with other substances which will be around it. These chemical properties are particularly important in some situations.

By observing the reactions between metals and air, water and dilute acid it is possible to arrange the metals in order of their reactivity.

The metal chosen for the drum of this washing machine must not react with the chemicals in washing powders

The metal chosen for these road lights must be able to resist attack by substances in the atmosphere for many years

The metals (there are two) chosen for this food can must not react with the foods inside

Magnesium, iron and copper reacting with hydrochloric acid

Metals with acid

The photograph shows some metals reacting with dilute hydrochloric acid.

The **effervescence** which you can see in the cases of magnesium and iron is caused by bubbles of hydrogen gas being formed as the reaction between the two substances proceeds. The other products of these reactions are the salts – magnesium chloride and iron chloride. (See page 36.)

$$\text{magnesium} + \frac{\text{hydrochloric}}{\text{acid}} \rightarrow \frac{\text{magnesium}}{\text{chloride}} + \text{hydrogen}$$
$$Mg(s) + 2HCl(aq) \rightarrow MgCl_2(aq) + H_2(g)$$

$$\text{iron} + \frac{\text{hydrochloric}}{\text{acid}} \rightarrow \frac{\text{iron}}{\text{chloride}} + \text{hydrogen}$$
$$Fe(s) + 2HCl(aq) \rightarrow FeCl_2(aq) + H_2(g)$$

Reactions with air (oxygen)

When metals react with oxygen they form oxides. Some metals burn brilliantly in oxygen. Magnesium, for example, is often used in distress flares because the brilliance of the light can be seen over a long distance.

The same reaction can be seen when magnesium ribbon is burned in the laboratory. (See photograph on page 11.) When the magnesium burns it reacts with the oxygen in the air to form magnesium oxide (a white powder).

$$\text{magnesium} + \text{oxygen} \rightarrow \text{magnesium oxide}$$
$$2Mg(s) + O_2(g) \rightarrow 2MgO(s)$$

Other metals, such as lead and copper do not burn in the same way. When heated in oxygen, a layer of their oxides forms on the surface of the metals. They are less reactive than magnesium.

Metals with water (steam)

Metals such as potassium, sodium and calcium react (violently) with cold water to produce an alkaline solution of the metal hydroxide. Hydrogen gas is released by the reaction.

The reaction of sodium with water, for example, produces sodium hydroxide and hydrogen.

sodium + water → sodium hydroxide + hydrogen
$$2Na_{(s)} + 2H_2O_{(l)} \rightarrow 2NaOH_{(aq)} + H_{2(g)}$$

Both sodium and potassium are so reactive that they have to be stored under oil to prevent them coming into contact with water or air.

However, because they have low melting points and are good conductors of heat they are used as coolants for nuclear reactors.

Other metals such as magnesium, zinc and iron react more slowly with water.

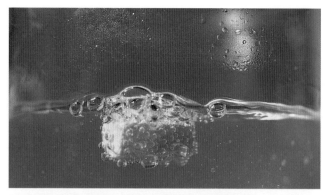
Sodium reacts quite vigorously with cold water

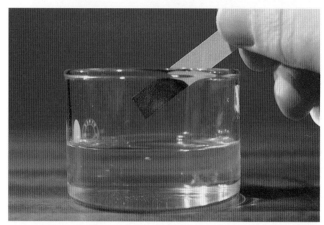
An alkaline solution is left after sodium reacts with water

Metals like potassium must be stored under oil

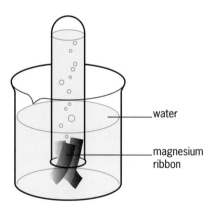
water

magnesium ribbon

Magnesium reacts slowly with water

A decorative iron gateway

A table of reactivity

If similar reactions are carried out using other metals then the metals can be arranged in order of their reactivity. This is known as a **reactivity series**. The table (page 15) shows the reactivity series for some common metals, with the most reactive at the top and the least reactive at the bottom.

Generally it is the relatively unreactive metals that we find the most uses for. Iron and copper, for example, can be found in many household and everyday objects.

The reactivity series for some common metals

Reactivity series	Reaction with dilute acid	Reaction with air/ oxygen	Reaction with water	Ease of extraction	
Potassium (K) Sodium (Na) Calcium (Ca) Magnesium (Mg) [Aluminium (Al)*] Zinc (Zn) Iron (Fe) Lead (Pb) Copper (Cu) Silver (Ag) Gold (Au) Platinum (Pt)	Produce hydrogen more and more slowly ↓ Do not react with dilute acid ↓	Burn very brightly and vigorously ↓ Burn to form an oxide with decreasing vigour ↓ React slowly to form the oxide ↓ Do not react ↓	Produce hydrogen more and more slowly with cold water ↓ React with steam with decreasing vigour ↓ Do not react with cold water or steam ↓	Difficult to extract ↓ Easier to extract ↓ Found as the element (native) ↓	I n c r e a s i n g R e a c t i v i t y ↑

*Because aluminium reacts so readily with the oxygen in the air a protective oxide layer is formed on its surface. This often prevents any further reaction and disguises aluminium's true reactivity. This gives us the use of a light and strong metal.

★ **THINGS TO DO**

1 For each use of a metal listed below, write a brief description of how its properties make it suitable for the job it does.
   a) Aluminium is used as window frames.
   b) Copper pans are used to make acidic foods such as jam and chutneys.
   c) Food cans are made from steel which is covered with a layer of tin.

2 Plan how you could carry out your own tests to find out the reactivity of five metals which your teacher will provide. You must think carefully about what you will be trying to test. If your teacher agrees to your plan, carry out your tests and prepare your own reactivity series for the metals you tested.

3 Write word and balanced chemical equations for the reactions between:
   a) calcium and oxygen;
   b) potassium and water;
   c) iron and dilute hydrochloric acid.

Using the reactivity series

The repair of railway lines, shown in the photograph, is carried out using a reaction known as the thermit reaction. A mixture of aluminium and iron(III) oxide is placed between the sections of track and ignited. A vigorous reaction takes place during which molten iron is formed. As the iron cools, it hardens effectively welding the sections of track together.

The reaction only takes place because the two metals involved – iron and aluminium – have different reactivities.

This repair is using the different reactivities of iron and aluminium

Competition for oxygen

If a metal high in the reactivity series is heated with the oxide of a less reactive metal, then the more reactive metal will take the oxygen away from it. You can think of it as the more reactive metal being 'stronger' and taking something away from the 'weaker' less reactive metal. In the thermit reaction, aluminium and iron(III) oxide react together.

The more reactive aluminium takes the oxygen from the less reactive iron. The reaction is an **exothermic reaction** – a reaction which releases energy in the form of heat. Enough heat is generated by the reaction to melt the resulting iron.

The reaction can be summarised by the equation below:

The oxygen is taken by the stronger (more reactive) metal

$$\text{aluminium} + \frac{\text{iron(III)}}{\text{oxide}} \rightarrow \text{iron} + \frac{\text{aluminium}}{\text{oxide}}$$

$$2Al_{(s)} + Fe_2O_{3(s)} \rightarrow 2Fe_{(s)} + Al_2O_{3(s)}$$

This type of 'oxygen capturing reaction' is also used to produce other metals from their oxides. Chromium, for example, is extracted from its oxide using aluminium.

Oxidation and reduction

When an element combines with oxygen it is oxidised. The process is called **oxidation**. A substance which loses its oxygen is reduced. The process is called **reduction**.

In the thermit reaction the aluminium has combined with oxygen to form aluminium

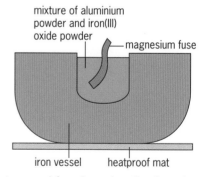

The apparatus used for observing the thermit reaction

oxide. The aluminium has been oxidised. The iron, however, has lost its oxygen and it has been reduced. The substance which accepts oxygen (and hence brings about reduction) is called the **reducing agent**, whilst the substance which gives the oxygen (and hence brings about oxidation) is called the **oxidising agent**.

Reduction and oxidation take place at the same time. This joint process is called a **redox** process.

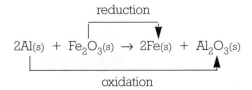

$$\text{2Al(s)} + \text{Fe}_2\text{O}_3\text{(s)} \rightarrow \text{2Fe(s)} + \text{Al}_2\text{O}_3\text{(s)}$$

reduction

oxidation

Displacement reactions

Small craft items, such as the silver Christmas tree shown in the photograph can be made by dipping copper foil into a solution of silver nitrate.

Some metals can displace or 'push' other metals out of their compounds. Copper, for example, is more reactive than silver, so it displaces, or pushes out, the silver from the silver nitrate solution. The silver atoms stick to the copper, covering it. The reaction is called a **displacement reaction**. The more reactive copper displaces the less reactive silver from the solution of silver nitrate.

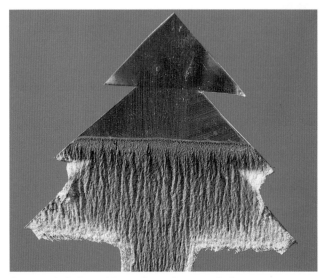

Copper foil dipped in a silver nitrate solution

The reactions of acids with metals are also examples of displacement reactions. In these reactions hydrogen is displaced.

★ THINGS TO DO

1 Try making one of the trees yourself.

Plan and carry out your own tests, under the supervision of your teacher, to find out how the mass of the silver which is deposited on the copper depends on the time for which the copper is left in the silver nitrate solution.

Prepare a report about what you find out and try to explain your findings.

2 One class did some tests to find out which other metals could be displaced from their solutions. Their results are shown opposite.

Use the pupils' results to answer the following questions in your notebook.

a) Why did they try both copper sulphate *and* copper nitrate?
b) Make a list of the metals in order of how well they reacted, with the most reactive at the top and the least reactive at the bottom.

Compare your list with the reactivity series on page 15 of this book. Make a note of any similarities you notice.
c) Make another list of the metal compounds with the most reactive at the top and the least reactive at the bottom. Compare this list with the reactivity series on page 15. Again make a note of any patterns which you notice.
d) How do you explain the difference between the lists produced in questions **b)** and **c)**?

The tick shows that there was a chemical reaction when we mixed the chemicals. A cross shows that there was no reaction.

metal used	copper	iron	lead	magnesium	silver	zinc
metal compound used in solution						
copper nitrate	x	✓	✓	✓	x	✓
copper sulphate	x	✓	✓	✓	x	✓
iron sulphate	x	x	x	✓	x	✓
lead nitrate	x	✓	x	✓	x	✓
magnesium sulphate	x	x	x	x	x	x
silver nitrate	✓	✓	✓	✓	x	✓
zinc sulphate	x	x	x	✓	x	x
score	1	4	3	6	0	5

Extraction of metals from their ores

We obtain most of the metals we use from the Earth's crust. The vast majority of metals are too reactive to exist as natural elements and are found as compounds in ores. These ores are usually oxides, carbonates or sulphides of the metal, mixed with impurities such as sand.

Extracting metals from their ores

Metals have been used for thousands of years. Our ancient ancestors used gold and silver as these were the only metals found which occurred naturally – they are so unreactive that they remain unaffected by the substances around them. Over hundreds of years people discovered that other metals could be obtained from their ores. The Bronze Age (from about 1800–600BC) is so called because during that period many weapons and tools were made from bronze.

Because bronze is a mixture of copper and tin, it is reasonable to assume that people must by that time have discovered how to extract copper and tin from their ores.

The Iron Age followed the Bronze Age from 600BC–43AD. This suggests that by that time people had discovered how to extract iron from its ore.

These metals were all easily extracted simply because they are low in the reactivity series and were probably first discovered by accident when a fire was surrounded by rocks containing the ores of these metals.

Iron is extracted from this ore – haematite

In the Bronze Age axe heads were made of bronze

The carbon formed from the burning wood and the heat produced would have been sufficient to reduce the ores in the surrounding rocks to their metals. Tests carried out using the compounds of other metals show that carbon can reduce the oxide of any metal below aluminium (in the reactivity series) to its metallic state.

The oxides of all metals below aluminium can be reduced by heating the oxides with carbon

potassium
sodium
calcium
magnesium
aluminium   ← carbon
zinc
iron
lead
copper
silver
gold
platinum

Extraction methods

The first step in the process of extracting a metal from its ore is usually the same and involves:

- crushing the ore with very heavy machinery;
- concentrating the ore by removing impurities;
- purifying the ore;
- converting the ore to the oxide of the metal.

There are four basic methods which are used to extract a metal from its ore:

- heating the ore;
- reduction with carbon;
- competition reactions;
- electrolysis.

Heating the ore

The lowest metals in the reactivity series (which are found in compounds in which the atoms are held together very weakly) can be extracted just by roasting their ores. Mercury, for example, can be extracted from its ore – cinnabar – by this method.

$$\text{mercury sulphide} \xrightarrow{\text{heat}} \text{mercury metal} + \text{sulphur dioxide}$$

$$HgS_{(s)} \xrightarrow{+O_2 \text{ from air}} Hg_{(l)} + SO_{2(g)}$$

Reduction with carbon

Compounds containing iron, copper and zinc have stronger bonds between their atoms. Carbon, however, attracts oxygen more strongly than any of these metals. When their oxides are heated with carbon, the oxides are reduced to form the metallic elements, whilst the carbon is oxidised.

$$\text{copper(II) oxide} + \text{carbon} \xrightarrow{\text{heat}} \text{copper} + \text{carbon dioxide}$$

$$2CuO_{(s)} + C_{(s)} \longrightarrow 2Cu_{(s)} + CO_{2(g)}$$

This is one of the cheapest methods of extracting metals from their compounds.

Competition reactions – when the ore is heated with a more reactive metal

More reactive metals must be obtained from their ores by reaction with an even more reactive metal in a **competition reaction** – a process which is more expensive than carbon reduction. Much of the high cost is because the metals which are used in the extraction process are themselves expensive to extract.

Titanium, for example, is extracted by heating titanium(IV) chloride (produced from its ore rutile) with the more reactive sodium metal.

$$\text{titanium(IV) chloride} + \text{sodium} \rightarrow \text{sodium chloride} + \text{titanium}$$

$$TiCl_{4(l)} + 4Na_{(l)} \rightarrow 4NaCl_{(s)} + Ti_{(l)}$$

Electrolysis

This (see page 28) is by far the most expensive form of extraction but is the only one which can be used for some of the more reactive metals. It is expensive because vast amounts of electricity are required to separate the atoms in the compounds. During **electrolysis** the purified ore is first melted and then electricity passed through it to separate the metal from other elements in the ore. Aluminium is a reactive metal separated from its ore by electrolysis.

★ THINGS TO DO

1 Use a hand lens to observe the ores of some naturally occurring metals. Prepare a key which would allow someone to identify each of the ores. You may also be able to do some other tests on the ores, such as how they react with dilute acids.

   Test your key on a friend. If necessary, take steps to improve it.

2 Write down word equations to describe how carbon reduction could be used to extract the metals lead and zinc from their oxides.

3 Use your research skills to find out how zinc is extracted from its ore.

Iron

Iron is the second most abundant metal in the Earth's crust. It is the most widely used metal although we usually use it in the form of its alloy steel rather than as pure iron. Over 400 million tonnes are produced worldwide each year.

Extraction of iron

Iron is extracted from its ores, haematite (Fe_2O_3) and magnetite (Fe_3O_4) in a blast furnace.

The blast furnace is a steel tower approximately 30 metres high, lined with heat resistant bricks. It is loaded with the 'charge' of iron ore (usually haematite), coke (made by heating coal in the absence of air), and limestone (calcium carbonate). A blast of hot air is sent in near the bottom of the furnace. This makes the 'charge' in the furnace glow, as the coke burns in the preheated air. This reaction produces carbon dioxide and as it does so gives out a lot of heat (exothermic reaction).

$$carbon + oxygen \rightarrow carbon\ dioxide$$
$$C_{(s)} + O_{2(g)} \rightarrow CO_{2(g)}$$

A number of chemical reactions then follow:

- the limestone begins to break down (thermal decomposition, see page 10) to form calcium oxide and carbon dioxide gas:

$$\begin{array}{ccc} calcium & calcium & carbon \\ carbonate & \rightarrow & oxide & + & dioxide \end{array}$$
$$CaCO_{3(s)} \rightarrow CaO_{(s)} + CO_{2(g)}$$

- the carbon dioxide gas from these two reactions then reacts with coke higher up in the furnace, producing carbon monoxide:

$$carbon\ dioxide + coke \rightarrow carbon\ monoxide$$
$$CO_{2(g)} + C_{(s)} \rightarrow 2CO_{(g)}$$

- the carbon monoxide is a reducing agent. It rises up the furnace and reduces the iron(III) oxide ore by taking the oxygen from it. This takes place at a temperature of approximately 700 °C. At this temperature the iron exists in its molten state.

Tapping off iron in a blast furnace

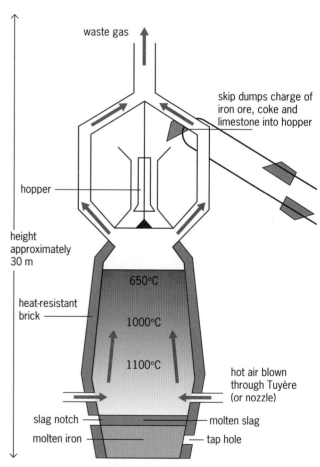

waste gas

skip dumps charge of iron ore, coke and limestone into hopper

hopper

height approximately 30 m

heat-resistant brick

650°C

1000°C

1100°C

hot air blown through Tuyère (or nozzle)

slag notch

molten slag

molten iron

tap hole

Cross section of a blast furnace

$$\begin{array}{ccc} iron(III) & carbon & carbon \\ oxide & + & monoxide & \rightarrow & iron + & dioxide \end{array}$$
$$Fe_2O_{3(s)} + 3CO_{(g)} \rightarrow 2Fe_{(l)} + 3CO_{2(g)}$$

The molten iron trickles down to the bottom of the furnace. The iron can be run out of the furnace through a hole in the base.

• the calcium oxide (formed by the decomposition of the limestone) reacts with any acidic impurities, such as silicon(IV) oxide (sand), in the iron ore to form a liquid slag which is mainly calcium silicate.

calcium oxide + silicon(IV) oxide → calcium silicate
$$CaO_{(s)} + SiO_{2(s)} \rightarrow CaSiO_{3(l)}$$

This material also trickles down to the bottom of the furnace, but because it is less dense than the molten iron, it floats on top of it. The molten slag is then run off at regular intervals and is used by builders and road makers for foundations.

The waste gases, mainly nitrogen and oxides of carbon, escape from the top of the furnace and are used to heat incoming air. This helps to reduce the energy costs of the process.

Blast furnaces operate 24 hours each day for several weeks before they must be closed down to replace the linings. The process is, however, much cheaper than methods such as electrolysis, used to extract aluminium and sodium from their ores.

The iron which is obtained from the blast furnace is known as pig or cast iron and contains about 4% carbon (as well as some other impurities). Because it is brittle and hard, the iron produced by this process has limited use. Gas cylinders are sometimes made of cast iron since they are unlikely to get deformed (bent) during their use. Most of the iron produced by the blast furnace is converted into different steel alloys.

Steel manufacture

Iron is converted into steel using the basic oxygen process. The basic oxygen furnace can convert up to 300 tonnes of iron to steel per hour.

• Calcium oxide (lime) is added to remove some of the impurities as slag.
• Magnesium powder is added to remove sulphur as magnesium sulphide (which mixes with any slag formed).
• Carbon is removed from the pig iron by blasts of oxygen gas which convert it to carbon dioxide gas. When the required amount of carbon has been reached, the oxygen is turned off. At this point other metals may be added to make different types of steel, such as, stainless steels.

Different types of steel have quite different properties which depend only on the amount of carbon, or other metals, which are mixed with the iron. The different properties affect the way the steels are used (see page 7).

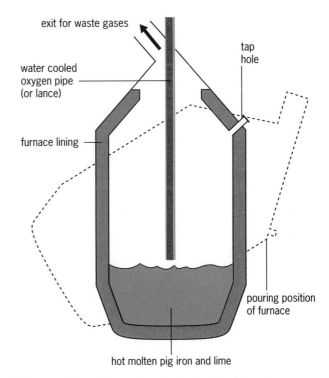

The vessel used to make steel, showing the water cooled lance

---

★ **THINGS TO DO**

1 Which raw materials are needed for the extraction of iron from its ore in a blast furnace?

2 a) Name two metallic elements which may be added to the basic oxygen furnace to produce different varieties of steel.

b) Give two advantages of stainless steel compared to cast iron.

3 Use the reactivity series to explain why other metals cannot be obtained from their ores in the same way.

An expensive oxidation process

Most metals begin to corrode after a time. **Corrosion** is due to a chemical reaction which occurs between the metal and other chemicals around it (its environment). Every metal in the reactivity series will corrode in some way. Generally, the higher the metal is in the reactivity series, the faster it will corrode. Iron and steel are the most commonly used metals with uses ranging from car bodies to artificial hip joints.

The corrosion of iron and steel is called **rusting**. Rust is an orange-red powder consisting mainly of hydrated iron(III) oxide ($Fe_2O_3.xH_2O$). Rusting is a serious problem. It is estimated that upwards of £500 million are spent each year replacing and repairing iron and steel structures.

Rust causes the steel or iron objects to become unattractive and may cause weaknesses in the structures they have been used to construct. Most manufacturers of iron and steel products take steps to prevent rusting. These steps can only be taken if we know what causes iron and steel to rust in the first place.

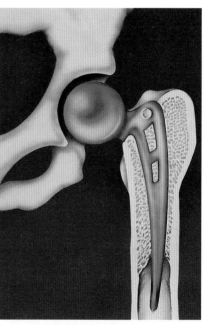

Cars are made of steel (*top left*).
A chemical plant (*bottom left*).
An artificial hip joint made from steel (*above*)

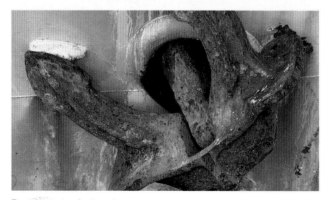

Rusting steel structures

Manufacturers take steps to prevent rusting

What affects the rate at which iron rusts?

The simple series of test tube reactions shown in the illustration allows us to identify the factors which cause iron to rust. After five days the following results had been recorded.

Tube	Conditions	Result
1	Control	No rust
2	Water and oxygen	Rust
3	Oxygen	No rust
4	Water	No rust
5	Salt, water and oxygen	Lots of rust

Both water and oxygen are essential for iron to rust and if either of these two substances is not present, then rusting will not take place. The rate at which an iron or steel object rusts is accelerated by the presence of salt, sodium chloride. This is a major problem for large objects such as piers and ships which are continually exposed to salt from the sea, and for cars and bridges which are exposed to salt during the winter months when it is spread on the roads to prevent the formation of ice.

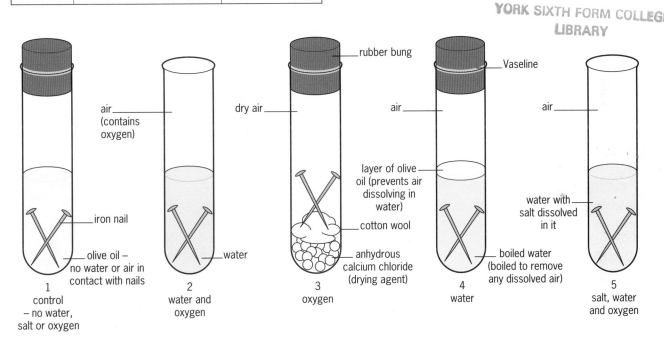

Experiment to find out what makes iron rust

★ THINGS TO DO

1 One way of coping with rust is to replace objects once made from iron or steel with other materials such as plastics. Make a list of some examples of where this has happened.

2 a) Car exhaust pipes corrode more rapidly than any other part of a car. This could be due to:
i) steel rusts more rapidly in the presence of acidic gases (such as those in the exhaust) or
ii) steel corrodes more rapidly in warm conditions.

Plan your own tests to find out whether these ideas are correct.

Carry out your tests when your plan has been approved by your teacher and prepare a report.
b) Motorists can now buy stainless steel exhausts, although they are much more expensive than 'ordinary' steel exhausts.

Write a plan describing how you could find out whether stainless steel lasts longer than 'ordinary' steel in the conditions of the exhaust pipe.

Rust prevention

To prevent iron rusting, we must stop oxygen (in the air) and water coming into contact with it. There are several ways to do this, all of which add to the cost of producing the goods.

Painting

Ships, lorries, cars, bridges and many other iron and steel structures are painted to prevent rusting. If the paint is scratched, the iron beneath it will start to rust and corrosion can then spread under the paintwork which may look unaffected. This is why it is important that paintwork is kept in good condition and checked regularly.

Oiling and greasing

The iron or steel in the moving parts of machinery is prevented from coming into contact with air or moisture by coating it with oil. This is the most common method of protecting moving parts of machinery, but the protective film must be renewed regularly. The oil and grease also help to reduce friction and wear as the parts move against one another.

Coating with plastic

Refrigerators, freezers and many other items are coated with plastic, such as PVC, to prevent the steel rusting. Again this method will only work if the plastic coating remains intact.

Plating

A thin layer of tin covers the steel used to make food cans. Because tin is less reactive than iron it does not corrode as quickly. If the tin layer is scratched then the iron underneath will rust.

Some taps, as well as bicycle handle-bars, are **electroplated** with nickel, then chromium to prevent rusting. The chromium also makes them look more attractive.

Some cars have more than 20 coats of paint. But, if the paintwork is damaged, then rusting will occur. This is because the water and oxygen can now react with the iron underneath

Oil and grease lubricate and protect machinery from rusting

These cans have been coated with tin

The handle-bars have been chromium plated

Galvanising

Some steel girders, used in the construction of bridges and buildings, coal bunkers and steel dustbins are **galvanised**. This involves dipping the object into molten zinc. When lifted out the steel object retains a thin surface coat of zinc. If the surface coat of zinc is damaged the iron still does not rust (see diagram). This is because zinc is more reactive than iron. The zinc will react in preference to iron to form zinc ions, Zn^{2+} (see page 107). Since the iron does not react it does not corrode.

Sacrificial protection

Bars of zinc are attached to the hulls of ships. Zinc is above iron in the reactivity series and will react in preference to it and so corrodes rather than the iron. As long as some of the zinc bars remain in contact with the hull, the ship will be protected from rusting. When the zinc bar has totally corroded it must be renewed. Gas and water pipes made of iron and steel are connected to blocks of magnesium for the same reason. In both cases, as the more reactive metal corrodes, it loses **electrons** (see page 108) to the iron and so protects it.

Some farm gates are galvanised to prevent rusting

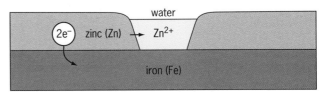

Zinc-coated iron does not rust

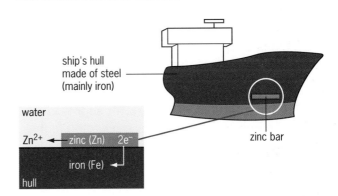

The zinc is sacrificed to protect the hull

★ THINGS TO DO

1 Rusting is another redox reaction. Explain the process of rusting in terms of oxidation and reduction.

2 What type of rust prevention method would be used to treat the following objects? Give a reason for your choice in each case:
   a) Lawn mower   **b)** Bicycle chain
   c) Garage door   **d)** Humber Bridge

3 Many chemicals are sold to treat rust. Some claim that they prevent rust recurring.
   Carry out some tests, under the supervision of your teacher, to find out which form of treatment is best. Prepare a report, using your data, which would help a car owner choose the best treatment from those you have tested.

Rust treatment and prevention

Not what they seem

Many metal objects which you see are made from iron or steel which has been coated with a thin layer of another, often less reactive metal. This may be done to make them more attractive and also to prevent corrosion.

The layers of metal are applied using a process called electroplating – an electric current is used to deposit atoms of the coating onto the steel base metal.

This watch has a thin coating of gold over steel

Silver plating teapots

Electroplating

Electroplating is the name of the process used to coat a base metal, such as steel, with another metal. During electroplating, the object to be coated is connected to the negative terminal of a power supply. A block of the metal which is to be coated onto the steel is connected to the positive terminal of the power supply. Together, these are called the **electrodes**. The electrode connected to the positive terminal of the power supply is called the **anode**. That which is connected to the negative terminal is called the **cathode**.

The electrodes are placed in a solution which contains atoms of the element with which the steel is to be coated. The solution must be an **electrolyte** – a solution which can conduct electricity. To silver plate the cutlery, the anode is a block of pure silver. The electrolyte is a solution of silver nitrate.

Separating elements from their compounds

Electrolysis can also be used to separate elements either from solutions of their compounds, or from their molten compounds. Only ionic compounds (see page 116) can be split up in this way. The main difference between this type of electrolysis and that of electroplating is that the electrodes are normally carbon. At the high temperatures at which the reaction takes place, the carbon rods slowly burn away.

Lead bromide, for example, is a white solid which does not conduct electricity. In its molten

When the circuit is closed, electrons are drawn from the silver atoms of the *anode*. This leaves behind positively charged particles of silver, called ions. The silver ions are positively charged (as are all metal ions). The positively charged silver ions leave the anode and pass into the solution.

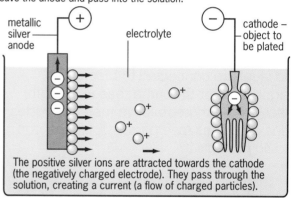

metallic silver anode

electrolyte

cathode – object to be plated

The positive silver ions are attracted towards the cathode (the negatively charged electrode). They pass through the solution, creating a current (a flow of charged particles).

The silver ions stick to the sides of the *cathode*. Electrons are transferred back to the silver ions, and they become silver atoms once again. The steel cutlery has now been coated with silver metal.

The process of silver plating can be thought of in three stages, although all take place at the same time

state, however, it becomes a conductor because the ions from which it is made become free to move around. (In the solid state the ions are closely bound together and are unable to move freely.)

In the molten state the lead bromide consists of positive lead ions and negative bromide ions, both of which are free to move. Because the lead ions contain two more protons than electrons (see page 114), they are written as Pb^{2+}. The bromide ions on the other hand, have one extra electron. They are written as Br^-. The illustration (page 27) shows what happens during the electrolysis of molten lead bromide.

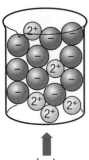

When molten the Pb^{2+} and Br^- ions are free to move and the molten material conducts electricity

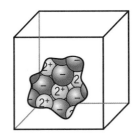

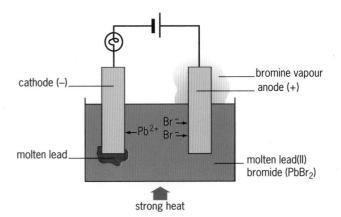

cathode (−)

bromine vapour

anode (+)

$\leftarrow Pb^{2+}$ Br^- →
Br^- →

molten lead

molten lead(II) bromide ($PbBr_2$)

strong heat

heat

The ions in this solid $PbBr_2$ are packed tightly together. Since they cannot move they cannot conduct electricity

Above The arrangement of the ions in solid $PbBr_2$ and the irregular arrangement in the molten state

Right During electrolysis the bromide and lead ions move towards the electrodes

The (positive) lead ions are attracted to the (negative) cathode. There they receive two negatively charged electrons. These balance out the excess positive charge on the ions, and they become (metallic) lead atoms.

lead ion + electrons → lead atom
Pb^{2+} + $2e^-$ → Pb

The gain of electrons which takes place at the cathode is called *reduction*.

The (negatively) charged bromide ions are attracted to the (positive) anode. To form bromine gas molecules each ion must first lose its negative charge at the anode and so form a neutral bromine atom.

bromide → bromine + electron
ion   atom
Br^- → Br + e^-

The loss of electrons which takes place at the anode is called *oxidation*.
Two bromine atoms then combine to form a bromine molecule:

two bromine → bromine
atoms   molecule
$2Br$ → Br_2

A handy way of remembering the rule for oxidation and reduction with electrons is OIL RIG (Oxidation Is Loss, Reduction Is Gain of electrons).

Overall, the reaction can be written as:

molten lead bromide → lead + bromine
$PbBr_{2(l)}$ → $Pb_{(l)}$ + $Br_{2(g)}$

★ THINGS TO DO

1 Explain the meaning of each of the following terms:
   a) electrolyte;   **d)** cathode;
   b) electrode;   **e)** reduction;
   c) anode;   **f)** oxidation.

2 Molten sodium chloride (salt) can be electrolysed in a similar way to lead bromide. Write down the names of the two products that would be formed and write word and symbol equations to represent the changes that would occur.

3 Any small steel or iron object can be copper plated using the apparatus shown.
Make a note of the factors which you think could affect the mass of copper deposited on

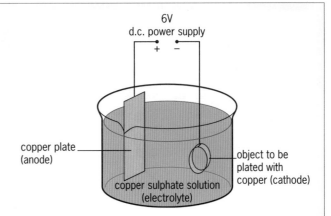

6V
d.c. power supply

copper plate (anode)

object to be plated with copper (cathode)

copper sulphate solution (electrolyte)

the cathode (the object you are plating). Give a reason for each one.
   Plan how you could carry out your tests. When you have carried them out, under the supervision of your teacher, prepare a detailed report and evaluation of your investigation.

Aluminium

Aluminium is a hard, silvery metal. Because it has a low density, objects made using aluminium are much lighter than similar ones made from steel. It is also a good conductor of heat and electricity. These properties make it especially useful in many situations.

Although it is a relatively reactive metal, when exposed to the air, it rapidly reacts with the oxygen forming a layer of aluminium oxide on the surface. This thin layer protects the aluminium from further corrosion.

The statue of Eros in Piccadilly Circus was cast from pure aluminium. When it was put up in 1893, the metal was still rare

Extracting aluminium

Aluminium is the second most abundant element in the Earth's crust but because it is so reactive, is only found in the form of aluminium compounds. Because it is so reactive, electrolysis is the only suitable means by which aluminium can be extracted from its ore – bauxite. (Aluminium is also found in other minerals such as cryolite and mica, and in much smaller quantities in clay.)

Pure aluminium oxide has a melting point of 2017 °C. Large amounts of energy would be needed to keep the aluminium oxide molten. In practice, the aluminium oxide is mixed with another mineral – cryolite. The melting point of the mixture is only 800–1000 °C, so significant amounts of energy are saved.

The molten aluminium oxide/cryolite mixture is electrolysed in a cell similar to that shown in the illustration (page 29). The anodes are blocks of graphite (carbon) which are lowered into the molten mixture from above. The cathode is the graphite lining of the steel vessel.

During electrolysis

Aluminium oxide is an ionic compound consisting of positively charged aluminium ions and negatively charged oxide ions.

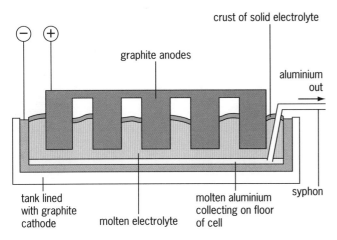

The Hall-Héroult cell is used to extract aluminium by electrolysis

When it is melted the ions become free to move as the strong forces of attraction between them are broken. During the electrolysis process the negatively charged oxide ions are attracted to the anode (the positive electrode). Here they lose electrons to form oxygen gas. The positive aluminium ions are attracted to the cathode (the negative electrode) where they gain electrons to form molten aluminium metal.

The overall reaction which takes place in the cell is:

$$\text{aluminium oxide} \xrightarrow{\text{electrolysis}} \text{aluminium} + \text{oxygen}$$
$$2Al_2O_{3(l)} \longrightarrow 4Al_{(l)} + 3O_{2(g)}$$

At the working temperature of the cell, the oxygen liberated reacts with the graphite (carbon) anodes producing carbon dioxide.

$$\text{carbon} + \text{oxygen} \rightarrow \text{carbon dioxide}$$
$$C_{(s)} + O_{2(g)} \rightarrow CO_{2(g)}$$

The anodes are burned away and must be replaced regularly. This is a continuous process in which vast amounts of electricity are used.

In order to make the process economic a cheap form of electricity is needed. Hydroelectricity is usually used to generate the electricity needed. The plant shown in the photograph uses the North Wales **hydroelectric power** scheme to provide some of the electrical energy it needs. Using cheap electrical energy has allowed aluminium to be produced in such large quantities that it makes it the second most widely used metal after iron.

There are some environmental problems associated with the siting of aluminium plants. These are:

- the extracted impurities form a red mud.
- the fine cryolite dust can be harmful. It is normally emitted through very tall chimneys to avoid contaminating the surrounding area.
- the claimed link between environmental aluminium (in the air and water) and a brain disease called Alzheimer's disease. It is thought that aluminium is a major influence in the early onset of this disease.

The aluminium smelting plant in Anglesey

★ THINGS TO DO

1 Produce a flow chart to summarise the processes involved in the extraction of aluminium metal.

2 List five uses of aluminium metal, stating which property (or properties) is (are) important for each use.

Copper

Copper is a fairly soft, salmon pink (when freshly made) unreactive metal. It is an excellent conductor of heat and electricity. These properties, allied with good ductility and malleability make it suitable for many uses around the home.

Most copper is found as the ore chalcopyrite ($CuFeS_2$). The ore is first converted to copper(I) sulphide and then roasted in air to give impure copper metal.

The copper used for electrical cables must be very pure. Any impurities reduce its electrical conductivity significantly by increasing its electrical resistance (see *GCSE Science Double Award Physics*, Topic 2.2). For that reason, it is purified using electrolysis.

In any street there may be several miles of copper wire in homes alone

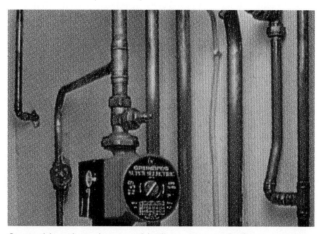

Central heating pipes and boilers are made from copper

Purification of copper by electrolysis

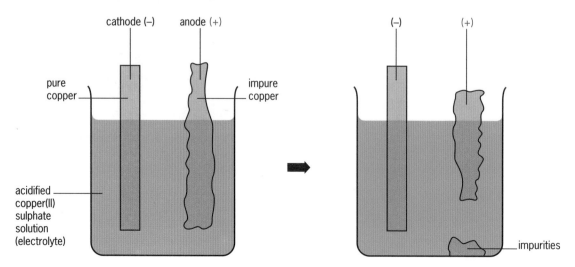

The copper purification process

The impure copper is used as the anode and is typically one metre square, 35–50mm thick and weighs 330kg. The cathode is a 1mm thick sheet weighing about 5kg, made from very pure copper. The electrolyte is a solution of copper(II) sulphate acidified with sulphuric acid.

When the current flows, copper ions move from the impure anode to the pure cathode. Any impurities fall to the bottom of the cell and collect below the anode in the form of a slime. This slime is rich in precious metals such as silver and gold and the recovery of these metals helps make the process more economic. The electrolysis proceeds for about three weeks until the anodes are reduced to about 10% of their original size, and cathodes weigh 100–120kg. The ions present in the solution are:

from the water: $H^+(aq)$, $OH^-(aq)$
from the copper(II) sulphate: $Cu^{2+}(aq)$, $SO_4^{2-}(aq)$

During the process the copper atoms lose electrons and pass into the solution as copper ions, $Cu^{2+}(aq)$, (see the diagram showing the movement of ions).

$$\text{copper atoms} \xrightarrow{\text{oxidation}} \text{copper ions} + \text{electrons}$$
$$Cu(s) \longrightarrow Cu^{2+}(aq) + 2e^-$$

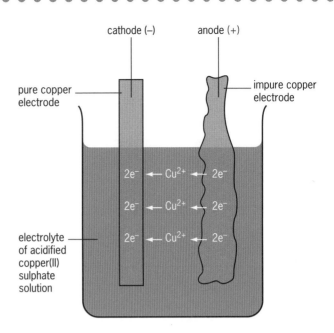

The movement of ions in the purification of copper by electrolysis

The electrons which are released at the anode travel around the external circuit to the cathode. There they are transferred back to the copper ions, $Cu^{2+}(aq)$, from the copper(II) sulphate solution forming copper atoms which are deposited on the cathode in the form of the metal.

$$\text{copper ions} + \text{electrons} \xrightarrow{\text{reduction}} \text{copper atoms}$$
$$Cu^{2+}(aq) + 2e^- \longrightarrow Cu(s)$$

★ **THINGS TO DO**

1 Why does copper have to be so pure to be used for electrical wiring?

2 List some other uses of copper.

3 Using the photograph as an example draw the apparatus you would use to copper plate a metal key.

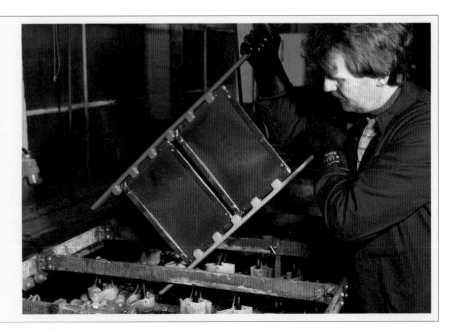

The object being plated is used as the cathode

pH and all that!

Acids and alkalis

If you have ever been stung by a bee or a wasp you will know of the 'burning' effect produced by **acids** and **alkalis**. Bees inject acids under the skin. The resulting 'itchy' effect is your body trying to neutralise their effects.

Acids are far more common than you perhaps realise. Many are found around the home in disguise.

Many fruits contain acids. The different acids produce the different flavours but they have one thing in common – they all taste sour.

The word acid means sour and all acids possess this property. They are also:

- soluble in water;
- corrosive;
- produce H^+ (hydrogen) ions in solution in water (aqueous solution).

Strong acid	Weak acid
hydrochloric acid	ethanoic acid
sulphuric acid	citric acid
nitric acid	carbonic acid (CO_2 dissolved in water)

Alkalis are the chemical opposite of acids. They:

- will remove the sharp taste from an acid;
- have a soapy feel;
- produce OH^- (hydroxide) ions in solution in water (aqueous solution).

Strong alkali	Weak alkali
sodium hydroxide	ammonia solution
potassium hydroxide	

Again, many alkalis are found around the home, often in cleaning fluids.

It would be far too dangerous to taste a liquid to find out if it was acidic! Instead we use substances called **indicators** which change colour when they are added to acids or alkalis.

What do all these foods have in common?

Apples contain malic acid, oranges and lemons contain citric acid, grapes contain tartaric acid and rhubarb contains oxalic acid

Air pollutants such as nitrogen oxides and sulphur dioxide dissolve in rainwater and form nitric and sulphuric acids. The acid rain harms and sometimes kills plants

Some common indicators

Indicator	Colour in acid solution	Colour in alkali solution
Phenolphthalein	Colourless	Pink
Methyl orange	Pink	Yellow
Red litmus	Red	Blue
Blue litmus	Red	Blue
Methyl red	Red	Yellow

Some common indicators, and the colour changes which take place with acids and alkalis, are shown in the table above.

To obtain an idea of how strong the acid or alkali is we must use another indicator known as **universal indicator**. The colour shown by this indicator can be matched against a **pH scale**. The pH scale runs from below 0 to 14. A substance with a pH less than 7 is an acid. One with a pH greater than 7 is alkaline and one with a pH of 7 is said to be neither acid nor alkaline, that is, **neutral**. Water is the most common example of a neutral substance. The illustration opposite shows the universal indicator colour range along with everyday substances which show the particular pH values.

Another way in which the pH of a substance can be measured is by using a pH meter as shown in the photograph. The pH electrode is placed into the unknown solution and its pH is shown on the digital display.

Some common alkaline substances

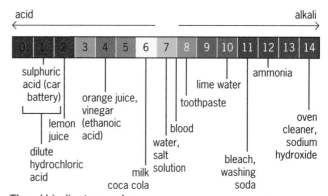
The pH indicator scale

A digital pH meter

★ THINGS TO DO

1 Prepare an information sheet for younger pupils describing how to use universal indicator to find the pH of solutions.

2 Acidic and alkaline solutions are formed when the oxides of elements are dissolved in water.

  Plan and carry out a series of tests, under supervision of your teacher, to find out whether there is any connection between the acidity or alkalinity of the solutions and the nature of the oxides, that is whether they are metal or non-metal oxides. Show your results clearly and write a general conclusion.

3 Gardeners need to know the acidity of their soil so that they can plan which plants will grow best.

  Plan how they could use universal indicator solution to test their soil.

  Write a leaflet for gardeners describing the steps they should take to test the pH of their soil.

4 Check the labels of a variety of foodstuffs. Make a list of the acids which are present in each.

Curing acidity

As you have seen in the previous topic, we come into contact with useful acids everyday. However some acids cause problems. For example, the tooth decay (dental caries) shown in the photograph is partly caused by the acids produced when bacteria break down sugar and other chemicals in food which is trapped between the teeth.

Excess acid in the stomach is a direct cause of indigestion. You normally treat it by taking an indigestion remedy which contains a substance which will react with and **neutralise** the excess acid.

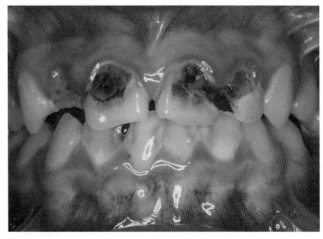

Tooth decay is caused partly by acid

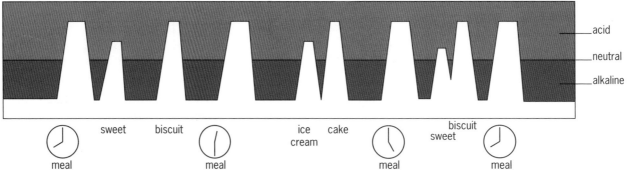

acid

neutral

alkaline

meal — sweet — biscuit — meal — ice cream — cake — meal — biscuit sweet — meal

The pH of your mouth changes like this during the day. It becomes acidic during and just after eating

Sometimes people get indigestion

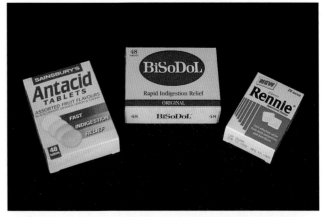

These tablets cure indigestion by neutralising stomach acid

In the laboratory if you wish to neutralise a common acid such as hydrochloric acid you can use an alkali such as sodium hydroxide. If you measure the pH of the acid as some sodium hydroxide solution is added to it, then the pH increases as the acidity is removed.

If equal volumes of the same concentration of hydrochloric acid and sodium hydroxide are added to one another, the resulting solution is found to have a pH of 7. The acid has been neutralised and a neutral solution has been formed.

$$\begin{array}{c} \text{hydrochloric} \\ \text{acid} \end{array} + \begin{array}{c} \text{sodium} \\ \text{hydroxide} \end{array} \rightarrow \begin{array}{c} \text{sodium} \\ \text{chloride} \end{array} + \text{water}$$

$$HCl_{(aq)} + NaOH_{(aq)} \rightarrow NaCl_{(aq)} + H_2O_{(l)}$$

When both hydrochloric acid and sodium hydroxide dissolve in water their ions separate completely. We may therefore write:

$$\begin{array}{c} H^+_{(aq)} \\ Cl^-_{(aq)} \end{array} + \begin{array}{c} Na^+_{(aq)} \\ OH^-_{(aq)} \end{array} \rightarrow \begin{array}{c} Na^+_{(aq)} \\ Cl^-_{(aq)} \end{array} + H_2O_{(l)}$$

You will notice that certain ions are unchanged on either side of the equation.

The $Cl^-_{(aq)}$ and $Na^+_{(aq)}$ ions are spectator ions

They are called spectator ions and are usually taken out of the equation because they are not involved in the reaction.

The equation now becomes:

$$H^+_{(aq)} + OH^-_{(aq)} \rightarrow H_2O_{(l)}$$

This type of equation is known as an **ionic equation**. The reaction between any acid and alkali in aqueous solution can be summarised by this ionic equation. It shows the ion which causes acidity (H^+(aq)) reacting with the ion which causes alkalinity (OH^-(aq)) to produce water (H_2O(l)) which is neutral. If you were to heat the solution in your beaker so that the water evaporated, the Na^+ and Cl^- would combine to form an ionic lattice. Sodium chloride would be left in the bottom of your beaker.

Sodium chloride is one of a range of compounds called **salts**. The formation of these substances will be discussed more fully in the next topic.

Other substances will also neutralise an acid. For example the substance in many of the indigestion remedies which reacts with and neutralises the excess acid is calcium carbonate. You will learn more about these substances in the next topic.

★ THINGS TO DO

1 Examine the diagram showing the change in the pH of your mouth during the day.
   a) What is the usual pH of your mouth?
   b) Why does the pH of your mouth fall after each meal?
   c) Toothpastes contain weak alkalis. How does cleaning your teeth regularly help prevent tooth decay?
   d) What other steps could be taken by you to prevent tooth decay?

2 Many lakes are affected by acid rain. In some, the high levels of acidity have killed all forms of life.
   Measures are now being taken to reduce the level of acidity by adding sodium carbonate to the water. The sodium carbonate neutralises the acidity of the water.

   Why do you think the amount of sodium carbonate which is needed will depend on the acidity of the water? Explain your answer using appropriate theory.
   Plan how you could test your hypothesis. If your teacher approves, carry out your tests and prepare a detailed report and evaluation.

3 **a)** Write both a chemical equation and ionic equation to represent the neutralisation of hydrochloric acid by potassium hydroxide.
   b) Write both a chemical equation and an ionic equation to represent the neutralisation of sulphuric acid by potassium hydroxide. Account for any difference you see between the ionic equations you have written.

Salts

In the previous section sodium chloride was produced when hydrochloric acid was neutralised by sodium hydroxide. Compounds formed in this way are known as salts. A salt is a compound which has been formed when all the hydrogen atoms of an acid have been replaced by metal atoms or by the ammonium ion (NH_4^+).

Uses of salts

Salt	Use
Silver bromide	In photography
Calcium carbonate	Extraction of iron, making cement, glass making
Sodium carbonate	Glass making, softening water, making modern washing powders
Ammonium chloride	In torch batteries
Calcium chloride	In the extraction of sodium, drying agent (anhydrous)
Sodium chloride	Making hydrochloric acid, for food flavouring, hospital saline, in the Solvay process for the manufacture of sodium carbonate
Tin(II) fluoride	Additive to toothpaste
Potassium nitrate	In fertiliser and gunpowder manufacture
Sodium stearate	In soap manufacture
Ammonium sulphate	In fertilisers
Calcium sulphate	For making plaster boards, plaster casts for injured limbs
Iron(II) sulphate	In 'iron' tablets

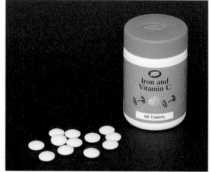

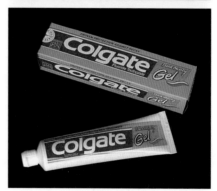

Salts are used in the manufacture of these items

Salts have many different uses as you can see from the table above.

If the acid neutralised is hydrochloric acid then salts called chlorides are formed. Other types of salts can be formed with other acids as shown in the table.

There are four general methods of preparing soluble salts. We will deal with two of these methods in this topic and the other two in the next topic.

Acid plus metal

This method can only be used with the less reactive metals. It would be very dangerous to use a reactive metal such as sodium in this

The different types of salts and the acids from which they have been formed

Acid	Type of salt	Example
Carbonic acid	Carbonates	Sodium carbonate (Na_2CO_3)
Ethanoic acid	Ethanoates	Sodium ethanoate (CH_3COONa)
Hydrochloric acid	Chlorides	Potassium chloride (KCl)
Nitric acid	Nitrates	Potassium nitrate (KNO_3)
Sulphuric acid	Sulphates	Sodium sulphate (Na_2SO_4)

type of reaction. The metals usually used for this method of preparation are the 'MAZIT' metals, that is, magnesium, aluminium, zinc, iron and tin. A typical experimental method is given below.

Excess magnesium ribbon is added to dilute sulphuric acid. During this addition an effervescence is observed due to the production of hydrogen gas.

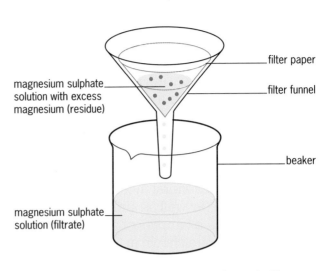

The excess magnesium (the residue) is filtered off

magnesium + $\dfrac{\text{sulphuric}}{\text{acid}}$ → $\dfrac{\text{magnesium}}{\text{sulphate}}$ + hydrogen

$$Mg_{(s)} + H_2SO_{4(aq)} \rightarrow MgSO_{4(aq)} + H_{2(g)}$$

The excess magnesium is removed by **filtration**. The magnesium sulphate solution is **evaporated** slowly to form a **saturated solution** of the salt.

The hot concentrated magnesium sulphate solution is tested by dipping a cold glass rod into it. If salt crystals form on the end of the rod the solution is ready to **crystallize** and so is left to cool. Any crystals produced on cooling are filtered and dried.

Acid plus carbonate or hydrogencarbonate

This method can be used with any metal carbonate and any acid providing the salt produced is soluble. The typical experimental procedure is similar to that carried out between an acid and a metal. For example, copper(II) carbonate would be added in excess to dilute nitric acid. There would be effervescence due to the production of carbon dioxide.

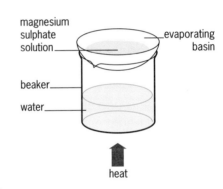

The solution of magnesium sulphate (the filtrate) is concentrated by slow evaporation

copper(II) carbonate + nitric acid → copper(II) nitrate + carbon dioxide + water

$$CuCO_{3(s)} + 2HNO_{3(aq)} \rightarrow Cu(NO_3)_{2(aq)} + CO_{2(g)} + H_2O_{(l)}$$

sodium hydrogencarbonate + nitric acid → sodium nitrate + carbon dioxide + water

$$NaHCO_{3(s)} + HNO_{3(aq)} \rightarrow NaNO_{3(aq)} + CO_{2(g)} + H_2O_{(l)}$$

★ THINGS TO DO

1 Copy and complete the following word equations and write balanced chemical equations for the following soluble salt preparations:
   a) magnesium + nitric acid →
   b) calcium carbonate + hydrochloric acid →
   c) zinc + sulphuric acid →

2 Devise an experiment to obtain copper(II) nitrate crystals from the salt solution produced at the end of the neutralisation reaction between copper(II) carbonate and nitric acid.

3 Describe the chemical tests you could use to identify:
   a) hydrogen gas;
   b) carbon dioxide gas.

4 Use your research skills to find out the uses of the following salts:
   a) magnesium sulphate;
   b) silver chloride;
   c) ammonium nitrate;
   d) copper(II) sulphate.

Bases

In this section we shall be looking at the other two methods of preparing soluble salts. Both these methods involve the use of **bases**.

A base is a substance which neutralises an acid producing a salt and water as the only products. If the base is soluble the term alkali can be used but there are many bases which are **insoluble**. In general most metal oxides and hydroxides (as well as ammonia solution) are bases. Some examples of soluble and insoluble bases are shown in the table.

Some soluble and insoluble bases

Soluble bases (alkalis)	Insoluble bases
Sodium hydroxide (NaOH)	Zinc oxide (ZnO)
Potassium hydroxide (KOH)	Copper(II) oxide (CuO)
Calcium hydroxide (Ca(OH)$_2$)	Lead(II) oxide (PbO)
Ammonia solution (NH$_3$(aq))	Magnesium oxide (MgO)

Acid plus alkali (soluble base)

This method is generally used for preparing the salts of very reactive metals such as potassium or sodium. It would certainly be too dangerous to add the metal directly to the acid. In this case we solve the problem indirectly and use an alkali which contains the particular reactive metal whose salt you wish to prepare. Because in this neutralisation reaction both reactants are in solution a special technique is required called a **titration**. Acid is slowly and carefully added to a measured volume of alkali using a burette, as shown in the photograph, until the indicator, usually phenolphthalein, changes colour. An indicator is used to show when the alkali has been neutralised completely by the acid. This is called the end-point. Now that you know where the end-point is you can add the same volume of acid to the measured volume of alkali but this time without the indicator. The solution which is produced can then be evaporated slowly to obtain the salt.

For example:

hydrochloric acid + sodium hydroxide → sodium chloride + water

$$HCl_{(aq)} + NaOH_{(aq)} \rightarrow NaCl_{(aq)} + H_2O_{(l)}$$

As previously discussed on page 35 this reaction can best be described by the ionic equation:

$$H^+_{(aq)} + OH^-_{(aq)} \rightarrow H_2O_{(l)}$$

A titration involves careful addition of the acid to the alkali until the indicator just changes colour

After slow evaporation the salt is left behind in the evaporating basin

Acid plus insoluble base

This method can be used to prepare salts of unreactive metals such as copper and lead. In these cases it is not possible to use a direct reaction of such a metal with an acid so the acid is neutralised using the particular metal oxide. The method is generally the same as that carried out between a metal carbonate and an acid, though some warming of the reactants may be necessary.

An example of such a reaction is the neutralisation of sulphuric acid by copper(II) oxide.

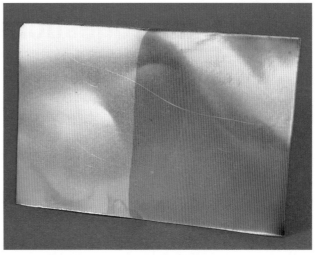

Citric acid in lemon juice has been used to clean up this piece of copper. It dissolves the oxide on the surface of the copper

$$\text{sulphuric acid} + \text{copper(II) oxide} \rightarrow \text{copper(II) sulphate} + \text{water}$$

$$H_2SO_4(aq) + CuO(s) \rightarrow CuSO_4(aq) + H_2O(l)$$

★ THINGS TO DO

1 With the aid of an example explain the meaning of the following terms:

  a) soluble base;
  b) insoluble base;
  c) titration;
  d) end-point.

2 Complete the following word equations as well as balanced chemical equations for the following soluble salt preparations:

  a) potassium hydroxide + nitric acid →
  b) zinc oxide + hydrochloric acid →
  c) ammonia + sulphuric acid →

3 a) Copy out and complete the table which is about the different methods of preparing salts.
  b) Write word and balanced chemical equations for each reaction shown in your table.

4 On an industrial scale it is important that waste is kept to a minimum. When preparing salts, for example, too much or too little of the reactants may mean that an excess remains when the reaction reaches its end-point – and that means that excess chemicals are wasted.

Starting with $25\,cm^3$ of potassium hydroxide solution, plan a series of class tests (under the supervision of your teacher) to find out how the amount of salt (potassium sulphate) which is obtained depends on the amount of sulphuric acid added to it.

When everyone has carried out their own test, collate the results and prepare your own report describing what you find out.

Method of preparation	Name of salt prepared	Two substances used in the preparation
Acid + alkali	Sodium sulphate	_____ and _____
Acid + metal	_____	_____ and dilute hydrochloric acid
Acid + insoluble base	Copper(II) sulphate	_____ and _____
Acid + carbonate	Magnesium _____	_____ and _____

Exam questions

denotes higher level questions

1 Two pupils added some dilute sulphuric acid to a series of metals. Some of their results are shown in the table opposite.

a) Complete the table by filling in the blank spaces. (2)

b) ONE of the results in the table is wrong.
i) Which result is wrong?
ii) Explain why it is wrong. (2)
c) How could the pupils identify the hydrogen gas given off? (2)
d) What would you expect to see if you added zinc metal to copper(II) sulphate solution? (2)
e) The metals copper and zinc form a substance called brass.

Two different samples of brass were made up as as shown in the table opposite.

Using this information, state whether brass is a compound or a mixture.
Explain your answer. (2)
(ULEAC, Specimen)

Metal	Gas given off	Other changes
Zinc	Hydrogen	Zinc reacts and disappears. Liquid stays colourless.
Magnesium	_____	Magnesium reacts and disappears. Liquid stays colourless.
Iron	No gas	Iron reacts and disappears. Liquid becomes pale green.
Copper	No gas	_____ _____ _____

	Copper	Zinc
Sample A	70%	30%
Sample B	85%	15%

2 The table opposite gives information about four metals **A**, **B**, **C** and **D**.

Answer the following questions, using only the information in the table.

a) i) Which metal is probably the most expensive to produce? (1)
ii) Which **two** metals would **not** be suitable for making household water pipes?
Give a reason in each case. (2)
b) When the ores of some metals are extracted from the ground there are large quantities of waste material. Why? (1)
(WJEC, 1993)

Metal	Reaction with moist air	Extraction from its ore	Other information
A	No reaction	Easy to extract	None
B	Reacts quickly	Easy to extract	A malleable metal
C	No reaction	Difficult to extract	None
D	No reaction	Easy to extract	Poisonous

3 A student wanted to find the position of tin in the reactivity series. He added small pieces of different metals to aqueous solutions of metal salts. The results are shown in the table below.

Solution	Metal added			
	iron	**lead**	**tin**	**zinc**
iron(II) sulphate	—	no reaction	no reaction	iron deposited
lead(II) nitrate	lead deposited	—	lead deposited	lead deposited
tin(II) chloride	tin deposited	no reaction	—	tin deposited
zinc(II) sulphate	no reaction	no reaction	no reaction	—

a) What is the name of the type of reaction that he used? (1)
b) Use the results to arrange the metals in order of **decreasing** reactivity. (1)
c) Write a word equation for the reaction between tin and lead nitrate solution. (1)
d) Use your knowledge of the reactivity series to explain how tin might be extracted from the ore cassiterite [tin(IV) oxide, SnO_2]. (3)
(ULEAC, 1995)

4 a) Wendy did an experiment to find out how reactive five different metals are.

She tested small amounts of each metal with cold water.

She tested small amounts of some of the metals with hot water and steam.

Some tests she did not try.

She wrote down what happened in the table opposite.

i) Arrange the metals **A**, **B**, **C**, **D** and **E** in order of decreasing reactivity. (1)

ii) Suggest why she did not try metal **C** in hot water or steam. (1)

b) Wendy has studied the reaction between some metals and dilute hydrochloric acid.

The metals react with the acid until they cannot be seen.

Bubbles of gas are produced during **all** the reactions.

She knows that

a metal + an acid → a salt + hydrogen

Write a word equation for the reaction of magnesium with dilute hydrochloric acid. (2)

(MEG, 1995)

Metal	Cold water	Hot water	Steam
A	no reaction	reacts slowly	burns if heated in steam
B	no reaction	no reaction	no reaction
C	reacts very vigorously	not tried	not tried
D	no reaction	no reaction	slow reaction
E	reacts slowly	rapid reaction	not tried

iv) Write a **word** equation for the reaction between segium and zinc chloride. (1)

b) i) Will segium react readily with cold water? Give a reason for your answer. (1)

ii) Will segium react readily if heated in steam? Give a reason for your answer. (1)

iii) **[A]** What gas will be made when segium is added to dilute hydrochloric acid? (1)

[B] Write down the name of the salt formed when segium is added to dilute hydrochloric acid. (1)

iv) Litmus is red in acidic solution, mauve (purple) in neutral and blue in alkaline solution.

Neutral litmus solution was added to some dilute hydrochloric acid. A large piece of segium was then added. What colour change would you see as a result of adding the segium? (1)

(SEG, 1994)

5 A metallic element called segium has been discovered. Like aluminium, it occurs in the Earth's crust as an oxide. Segium can be extracted from this oxide by heating it with aluminium powder. The word equation for this reaction is:

aluminium + $\dfrac{\text{segium}}{\text{oxide}}$ → $\dfrac{\text{aluminium}}{\text{oxide}}$ + segium

When a piece of segium metal is dropped into zinc chloride solution, the segium dissolves and crystals of zinc metal slowly form on the segium.

Some metals are listed below in order of how reactive they are.

Sodium (most reactive)

Calcium

Magnesium

Aluminium

Zinc

Iron

Lead

Copper (least reactive)

a) i) What name is given to this list of metals: activity series, decay series, ferrous series or periodic series? (1)

ii) Add an arrow labelled **"segium"** to show where in the list of metals above, segium would be placed. (1)

iii) **Explain** why you chose the position for your arrow. (2)

6 The diagram below shows a wire fence. The fence is held up by steel posts. The posts were painted each year. The fence was blown down by strong winds. The steel posts had rusted badly at ground level.

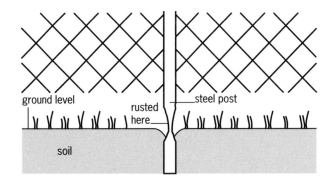

a) What **two** substances cause the steel to rust? (2)

b) How does painting help to stop rusting? (1)

c) Suggest why the steel posts had rusted much more at ground level. (2)

d) Suggest **one** other way that could be used to protect steel posts from rusting. (1)

(NEAB, 1995)

7 a) Look at the list of words.

acid   alkali   indicator   neutral

Put each word in its correct place in the sentences below. (3)
Hydrochloric .... reacts with an .... to make a salt and water. Universal .... can be used to show when the solution is ....
b) The pH scale is used to measure acidity.

0	1	2	3	4	5	6	7	8	9	10	11	12	13	14

Acidic ◄─────────── ───────────► Alkaline

i) Which pH number is neutral? (1)
ii) Universal indicator is used to find the pH of solutions. In each case, write down if the solution is acidic, alkaline or neutral. (2)

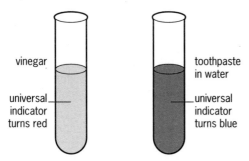

vinegar — universal indicator turns red

toothpaste in water — universal indicator turns blue

iii) Our stomach contains a very acidic solution. Suggest the pH number of the acid in the stomach. (1)

iv) When our stomach is too acidic we get stomach-ache. We can cure stomach-ache by taking stomach tablets.
How does the pH in the stomach change after taking the tablets? Explain your answer. (3)
(SEG, 1995)

8 The salt sodium hydrogen phosphate, (Na_2HPO_4), is used as a softening agent in processed cheese.

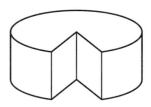

The salt can be made by reacting phosphoric acid (H_3PO_4) with an alkali.
a) Complete the name of an alkali that could react with phosphoric acid to make sodium hydrogen phosphate.
.... hydroxide (1)
b) What name is given to a reaction in which an acid reacts with an alkali to make a salt? (2)
c) Use the **Data Book** to answer these questions.
i) What colour is universal indicator in pure water? (1)
ii) A solution of phosphoric acid was tested with universal indicator solution. The indicator changed colour to orange.
What was the pH of the phosphoric acid solution? (1)
d) How would the pH change when alkali is added to the phosphoric acid solution? (1)
e) i) What ions are present when any acid is dissolved in water? (1)
ii) What ions are present when any alkali is dissolved in water? (1)
iii) Write a chemical equation for the reaction that takes place between the ions named in i) and ii). (1)
(NEAB, Specimen)

2

EARTH MATERIALS

Natural materials

We use millions of different materials every day. Some are found naturally on or under the Earth's surface. They are naturally occurring materials and have been used for thousands of years in many different ways.

Many of the millions of substances found in the Earth's crust are of little use in their natural form. We have found ways to change them into substances which are much more useful using chemical reactions.

Today, most of the materials we use are **manufactured** from naturally occurring compounds. Some of the copper we use, for example, is extracted from malachite, a naturally occurring mineral found in the Earth's crust. Minerals containing metals are called **ores**. Malachite is the raw material from which the copper metal is made. The Earth and the atmosphere provide the raw materials for almost everything we need.

Gold is one of the few metals found naturally. This death mask was found in the tomb of the Egyptian child King Tutankhamun – buried 3300 years ago

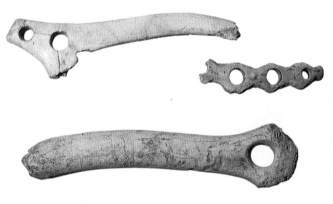

Our ancestors used stone to form weapons and tools. Later stone was used to provide shelter

Wood was easily fashioned to form houses and ships such as the Mary Rose, which sank in 1545

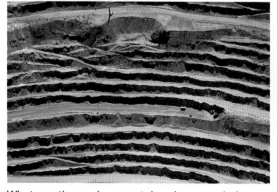

What are the environmental and economic issues concerned with mining copper on this scale?

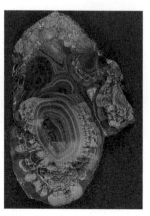

Malachite (copper ore)

Copper products

Plants are the raw materials from which we obtain timber, foods such as cereals, vegetables and fruit, cooking oil, clothing (from cotton or linen), rubber and many medicines and drugs. Growing plants absorb carbon dioxide from the atmosphere, replacing it with the oxygen we need to breathe. Carbon and other compounds are stored in the plants, providing the chemicals which animals need for survival.

The remains of plants and animals form **fossil fuels** such as coal, oil and gas. Fossil fuels have a high carbon content. When burned, they release the energy needed for our homes, offices and industries. They are also used to generate over 50% of the electricity we need. Although they have taken 300 million years to form they are burned in a few minutes.

The same fossil fuels are the raw materials from which we obtain hundreds of other products such as coke, coal gas, petrol, diesel and aviation fuels, dyes, plastics, perfumes, pesticide and detergents.

The sea provides us with some food. It is also the raw material from which important elements such as chlorine, iodine and magnesium are obtained. Most of the water vapour in the atmosphere evaporates from the seas and oceans around the world. It is carried across the surface of land, falling as rain and providing water for plants. The sea also absorbs some of the carbon dioxide from the atmosphere, forming compounds which fall to the sea bed and form new rocks.

We may think of the air as only providing the oxygen we need. It is also the raw material from which we obtain important gases such as oxygen, nitrogen and argon.

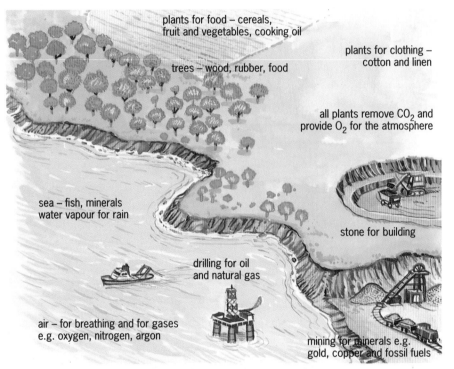

plants for food – cereals, fruit and vegetables, cooking oil

trees – wood, rubber, food

plants for clothing – cotton and linen

all plants remove CO_2 and provide O_2 for the atmosphere

sea – fish, minerals water vapour for rain

stone for building

drilling for oil and natural gas

air – for breathing and for gases e.g. oxygen, nitrogen, argon

mining for minerals e.g. gold, copper and fossil fuels

Naturally occurring materials are found on and under the Earth's surface. These are the raw materials from which things can be manufactured

---

★ THINGS TO DO

1 Make a list of objects which you find around your home. Say which materials are used to make them and what special properties those materials have which make them so useful.

2 Sort your list of objects into two groups – those which are made from naturally occurring materials, such as wood, and those which are made from materials, such as plastic, which are manufactured from raw materials. You may need to create a third group for objects which have both types of material in them.

3 Choose three of the objects in your list which are made from manufactured materials. Could naturally occurring materials be used to make them? If so, which materials would you use? What would be the advantages and disadvantages of using only natural materials for those things?

Limestone – an essential mineral

Limestone is a naturally occurring form of calcium carbonate ($CaCO_3$). Calcium carbonate is also found naturally as chalk, calcite and marble. It is the second most abundant mineral in the Earth's crust. Deposits of chalk were formed from the shells of dead marine creatures that lived many millions of years ago. In several places in Great Britain, the chalk was covered with other types of rock and was crushed under great pressure. This caused the relatively soft chalk to be changed into the harder material, limestone. Marble, one of the other major forms of calcium carbonate, was formed in places where the chalk was not only subjected to a high pressure but also a high temperature.

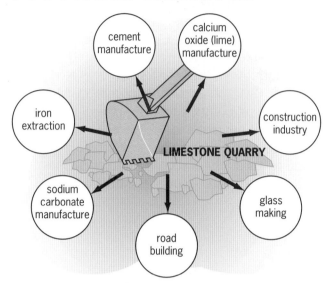

The uses of limestone

Uses of limestone

Limestone has many varied uses.

Neutralising acid soil

Powdered limestone is most often used here because it is cheaper than lime (calcium oxide) which has to be produced by heating limestone. The reaction which occurs between the limestone and the acid in the soil is a neutralisation process.

Manufacture of iron and steel

In a blast furnace limestone is used to remove earthy and sandy materials found in the iron ore to form a liquid slag which can be easily removed. More details of the extraction of iron and its conversion into steel are given in Topic 1.7.

Manufacture of cement and concrete

Limestone (or chalk) is mixed with clay (or shale) in a heated rotary kiln, using coal or oil as the fuel. The material produced is called cement. The dry product is ground to a powder and then a little calcium sulphate ($CaSO_4$) is added to slow down the setting rate of the cement. When water is added to the mixture complex chemical changes occur forming a hard interlocking mass of crystals.

The spreading of limestone onto soil to cure soil acidity

A modern rotary kiln for making cement

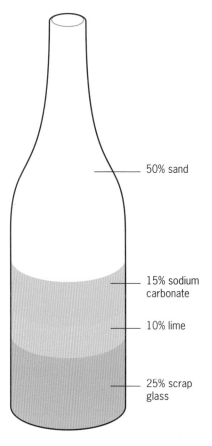

Left Reinforced concrete is used in the construction of tall buildings

Below The composition of glass

50% sand

15% sodium carbonate

10% lime

25% scrap glass

Concrete is a mixture of cement with stone chippings and gravel which gives it body. After the dry mixture has been mixed with water it is poured into wooden moulds and allowed to set hard. Reinforced concrete is made by allowing concrete to set around steel rods under tension. This gives it the greater **tensile strength** which is required for the construction of large bridges and tall buildings.

Lime manufacture

After further processing limestone is used for lime manufacture. When calcium carbonate is heated strongly it thermally decomposes to form calcium oxide and carbon dioxide.

$$\text{calcium carbonate} \rightleftharpoons \text{calcium oxide} + \text{carbon dioxide}$$

$$CaCO_3(s) \rightleftharpoons CaO(s) + CO_2(g)$$

This reaction is an important industrial process and takes place in a lime kiln. The calcium oxide produced from this process is known as quicklime or **lime**.

Calcium oxide (CaO) is a base (a base is a substance that neutralises acids) and it is still used by some farmers to spread on fields to neutralise soil acidity and to improve drainage of water through soils which contain large amounts of clay. It also has a use as a drying agent in industry. Another use for lime is in the manufacture of soda glass which is made by heating sand with soda (sodium carbonate, Na_2CO_3) and lime.

Large amounts of calcium oxide are also converted into calcium hydroxide ($Ca(OH)_2$) which is called slaked lime.

Manufacture of calcium hydroxide – slaked lime

Calcium hydroxide ($Ca(OH)_2$) is manufactured from calcium oxide (lime) which is made from limestone.

Calcium hydroxide, a white powder, is produced by adding an equal amount of water to calcium oxide. This process can be shown on the small scale in the laboratory by heating a lump of limestone very strongly to convert it to calcium oxide. Water can then be carefully added one drop at a time, to the calcium oxide. The reaction which occurs produces a great deal of heat. This process is called **slaking** and produces calcium hydroxide.

Calcium hydroxide is a cheap industrial alkali. It is used in large quantities to make bleaching powder, by some farmers to reduce soil acidity, in the manufacture of whitewash, glass manufacture and in water purification.

Calcium hydroxide (slaked lime) is also mixed with sand to give mortar. When mixed with water and allowed to set a strongly bonded material is formed which is used to hold bricks together.

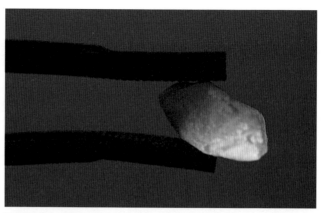
Heating limestone produces lime

Adding water to lime gives slaked lime

★ THINGS TO DO

1 Vast quantities of limestone are quarried each year to meet our needs. Unfortunately most of the purest limestone occurs in areas of great natural beauty, such as the Peak District. It has been estimated that if **quarrying** continues at its present rate, a quarter of all of the Peak District will be removed within the lifetime of somebody who is a child today.

Quarrying for limestone in the Peak District

Quarrying limestone and other materials can cause considerable damage to the environment, but the benefits we obtain from their use as raw materials must be balanced against the harm done to the surroundings.

Imagine you live in a village near to a large quarry. The quarry employs 200 local people. Every day 100 lorries pass through the village carrying materials to the railway. The quarry produces a lot of dust, and in wet weather the lorries leave a lot of mud on the roads. Many weekend visitors visit the area, which has a picnic site close by. They often stop off in the village to buy sweets, ices and other goods.

The quarry company have applied to double the size of the quarry. They argue that this is necessary to meet the demand for limestone. The company have said they will make sure the wheels of the wagons are cleaned before they leave the site, and they will do whatever they

★ THINGS TO DO (continued)

can to reduce dust. They say they will landscape the area around the quarry so that it cannot be seen from the road.

An environment group argue that the planning application should be refused. They say that the present quarry can supply enough limestone to meet industry's needs. They also say that if the quarry is extended, the habitat of many rare plants and animals will be destroyed. The dynamiting and constant traffic will also prevent the animals settling in new habitats nearby.

The trade unions argue that the extension will provide extra jobs, which will bring extra money into the local shops. The local council will benefit from increased rates. If the application is not allowed, the jobs of the men in industries which rely on the limestone may be lost.

Discuss in your class whether quarrying on such a huge scale should be allowed. You could organise a debate with groups making out arguments for and against the planning proposal by the company.

Make a list of the advantages and disadvantages of having quarries such as this. Try to include as many ideas as possible, such as the importance of the products obtained from limestone, the effect on jobs in the area, the effects on the environment, the effect on the roads, and so on.

Make a note of what your class decides would be the best way of tackling problems such as this.

2 Some foods are produced in 'self heating cans'. Quicklime is placed between the two sides of the can. When water is added it reacts with the quicklime and the food is heated.

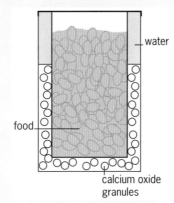
food
water
calcium oxide granules

Some pupils tested this heating method by placing a small can inside a larger one as shown in the illustration.

They added water to the quicklime and measured the temperature rise of some water placed inside the can.

These were their results:

	temperature in deg.C				
	0	2	4	6	8
			(minutes)		
large lumps of quicklime	15	15	17	21	26
smaller lumps of quicklime	15	16	23	29	34
powdered quicklime	16	23	32	41	46

a) Write a set of instructions which other people could use to do this test. Don't forget to include any safety instructions.

b) What steps would the children have taken to make their tests fair?

c) What do you think happens when water is added to the quicklime?

d) Which of the tests released heat most quickly?

e) If heat is released too quickly the food near the sides of the can will get very hot and may burn. If heat is released too slowly, then most of the heat could be lost through the sides of the can and the food may not warm. What would be the best size particles to use? Describe how you use the results to get your answer.

3 Devise an experiment that you could carry out in the laboratory to find out whether powdered limestone is better at curing soil acidity than calcium oxide.

The changing Earth

This island was formed as a result of volcanic activity

Over millions of years the surface of the Earth has slowly changed. The surface as we see it today has been created by:

- movements in the Earth's crust;
- glaciers;
- the weather;
- human activity.

In the earliest times, volcanoes (which were then almost 'flat') covered much of the land. Molten rock rose to the surface forming new land. Volcanoes erupting under the sea formed new land. The islands of Hawaii, for example, are a chain of volcanoes which rose from beneath the sea. Some are still active.

Huge mountain ranges were formed as the Earth's crust was crumpled by movements in the crust. **Fossils** of sea animals can be found several thousand feet above sea level (such as those on Mount Everest). This tells us that this land was once below sea level.

The Himalayan mountains

Forces deep within the ground push, pull, squeeze and bend the layers of rock under the surface, crumpling their shape and changing the arrangement of the particles in the rock. Movements such as this bring some of the compounds we need closer to the surface.

In some places the Earth's crust is fractured. The parts of the crust on each side of the fracture sometimes slide past one another. One side may slip down below the other. Rock movements such as this are called **faults**. **Rift valleys** are formed when pulling forces in the Earth's crust allow the block of the Earth's crust in the centre to slip downwards leaving raised sections on each side.

In places the shape of the land was formed by huge **glaciers** which slowly moved across the surface grinding and scraping it away (eroding it), leaving U-shaped valleys.

Folded rock at Lulworth Cove, Dorset

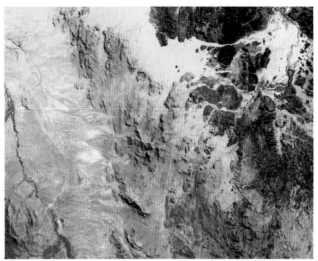

The Rift Valley in Africa from space

A U-shaped valley formed by glaciers, in Switzerland

The Grand Canyon, Arizona, USA

As water flows across the surface of the land it wears away the rock beneath it. Thousands of years of erosion such as this have left deep **canyons** and valleys.

Temperature changes slowly break down (**weather**) the surface of rock into smaller and smaller particles.

Slow processes

Most changes which take place at the surface of the Earth are so slow they are not noticed. Over thousands of years, however, the changes become very apparent.

A scree slope in the Dolomites

About 50 million years ago, for example, Africa moved slowly towards Europe. The rocks under Europe were squashed and folded by tremendous forces, forming the mountains we call the Alps.

The same force caused the land under Southern England to fold like this.

Slowly the folded rocks have been weathered and eroded. Now the same area looks like this.

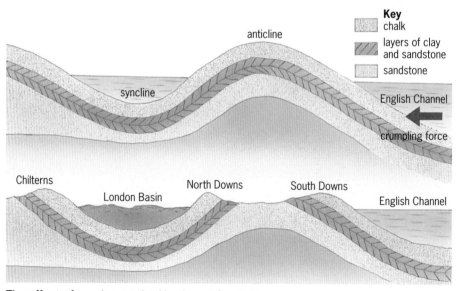

Key
chalk
layers of clay and sandstone
sandstone

anticline

syncline

English Channel

crumpling force

Chilterns

London Basin

North Downs

South Downs

English Channel

The effect of erosion on the North and South Downs

★ THINGS TO DO

1 a) Describe how weathering, **erosion**, transport and **deposition** have worked to change the appearance of the North and South Downs. (You will need to think about the properties of the rocks in the illustration of the North and South Downs.)
b) Would the present shape of the land have been different if the rocks had been harder? Explain your answer.
c) Draw a picture showing what the area might look like 10 000 years from now.

2 Write a description of your favourite place in the countryside or on the coast. It could be a place near your home, or somewhere you have visited whilst on holiday. Draw a picture of the area.

Describe any special features of the land, such as hills or mountains, rivers, streams, valleys and so on. Add some notes saying how you think they were formed.
Draw what you think the area will look like 1000 years from now.

3 Cairngorm is one of the highest mountains (1245 metres above sea level) in Scotland. Slowly, like all other mountains, Cairngorm is being worn away by weathering and erosion, although some steps are being taken to reduce the effects. The photographs show different areas of Cairngorm.
Talk with others in your group about what each photograph shows. Make a note of the ways in which the land is being changed.

Journey to the centre of the Earth

Many years ago Jules Verne, in his book *Journey to the centre of the Earth*, described how adventurers tunnelled to the centre of the Earth meeting large creatures who made their homes in enormous caverns. We now know that this type of journey would be impossible – unless we could drill a tunnel 6380 kilometres long through solid and molten rock at temperatures of several thousand degrees.

The Earth is not solid. Evidence gathered from waves sent out by earthquakes (see *GCSE Science Double Award Physics*, Topic 4.17) suggests that it has several layers, each with quite different properties.

The centre of the Earth, called the **inner core**, is a ball of very hot, dense material. Its density (17 g/cm^3), suggests it is largely iron and nickel. The temperature is over 4300 °C but the core stays solid because the pressure is so high.

The **outer core** is probably iron and nickel, with an average density of 12 g/cm^3. Even though the temperature is about the same as the inner core, the outer core is liquid because it is under less pressure.

The **mantle** is made of cooler, solid rock, although the upper layers of the mantle flow slowly (rather like thick toffee). The average thickness of the mantle is 2800 kilometres.

The **crust** is the outer layer and is made of a thin layer of solid rock. Below the oceans the crust is quite thin – only 5–10 kilometres thick. Under the continents it is thicker, varying between 30 and 90 kilometres thick.

90% of the Earth's crust is made up of compounds of oxygen, silicon, aluminium, iron, calcium, sodium, potassium and magnesium. Thousands of other compounds exist in much smaller quantities.

Central heating

As you go deeper into the Earth the temperature rises. Heat is also lost from the surface of the Earth by **radiation** every day (see *GCSE Science Double Award Physics*, Topic 4.14 for more information on radiation), so we would expect the Earth to cool down. This does not happen because heat is produced to make up for the heat which is lost. One reason may be that deep inside the Earth – in the core – the decay of **radioactive isotopes** releases huge amounts of energy (rather like the processes inside a nuclear reactor) which help keep the Earth warm. (See *GCSE Science Double Award Physics*, Topic 4.15.)

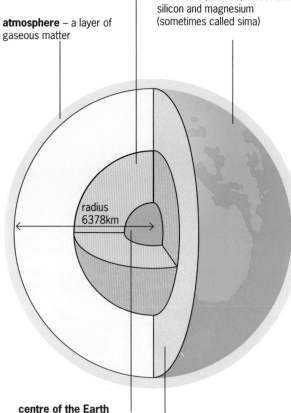

outer core – very dense liquid rock at high temperature composed of nickel and iron. The Earth's magnetic field arises here
density 10–12 g cm^{-3}
extends to a diameter of 6930 km

atmosphere – a layer of gaseous matter

crust – a 50km thick layer of solid rock (very thin compared with the diameter of the whole Earth) density 2.0–3.0 g cm^{-3}. Crust may be continental crust (granitic crust) rich in silicon and aluminium (sometimes called sial), or oceanic (basaltic) crust rich in silicon and magnesium (sometimes called sima)

radius 6378km

centre of the Earth (inner core) – solid rock at very high temperature and pressure composed of nickel and iron
density 12–18 g cm^{-3}
extends to a diameter of 2530km

mantle – a thick layer of solid, dense rock rich in magnesium and silicon
density 3.4–5.5 g cm^{-3}
parts of the mantle move slowly

The Earth

The crust is floating and cracked

Rocks in the Earth's crust are less dense than those below them and so they float on the partially molten mantle beneath them. There are two types of crust – the crust under the continents, called the **continental crust**, and the crust under the oceans, called the **oceanic crust**.

The continental crust contains lighter rocks such as granite, whereas the oceanic crust contains denser rocks such as basalt.

The outer layer of the Earth – the crust – is rather like a cracked egg shell. It is not one piece of rock, but several huge slabs called **tectonic plates**. The plates float (like rafts) on the part-liquid mantle, moving by as much as five centimetres each year. The part of the crust on which Great Britain rests was once very close to the Equator. Over millions of years it has moved slowly northwards to its present position.

Some evidence suggests that the continents, which are now separated by thousands of miles of ocean, were once all joined together in one huge land mass called Pangaea. If, for example, you look at the shape of South America and the west coast of Africa, they seem to fit together quite nicely.

Of course, this may be coincidence. But other evidence, obtained from the rocks and fossils on each of the continents, strongly supports the idea that they were once part of the same land mass.

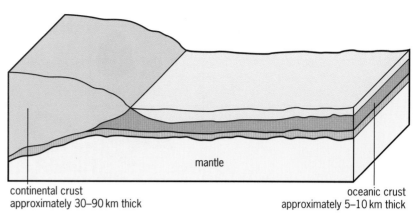

continental crust
approximately 30–90 km thick

oceanic crust
approximately 5–10 km thick

Continental and oceanic crusts

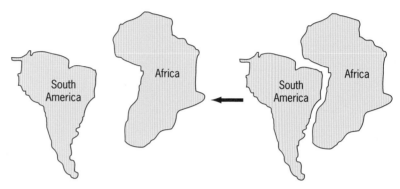

South America and Africa may have been joined together

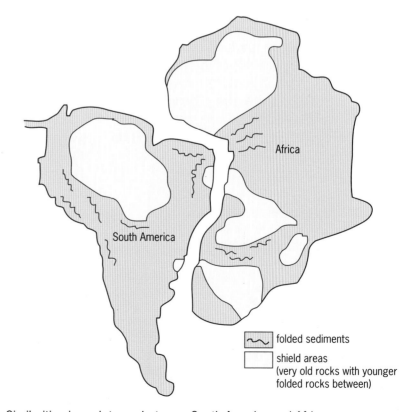

folded sediments

shield areas
(very old rocks with younger folded rocks between)

Similarities in rock types between South America and Africa

Time (millions of years ago)	South America	Africa
300	Glossopteris	Glossopteris
250	Mesosaurus	Mesosaurus
100	Tyrannosaurus	Allosaurus
50	No fossils of any type of horse found	Early types of horse
Present day	Llama	Camel

Looking at fossils to decide when the continents began to drift apart

Within the rocks on each of the 'modern' continents there are fossils of the same plants and animals – found in rocks which are of the same type and the same age. These animals must have been present before the land began to separate. By studying the fossils on each continent, it is also possible to identify when the plates began to drift apart.

Moving continents

The continents move because heat from the centre of the Earth generates strong **convection currents** in the mantle. (See *GCSE Science Double Award Physics*, Topic 1.10 for more information on convection currents.) These convection currents squeeze molten rock through any weaker parts of the crust. As the **magma** rises up through the crust, it continuously forces the two plates a little further apart – by as much as 5 cm each year. As the magma rises, it solidifies to form new rock. This can happen beneath the continents or beneath the oceans.

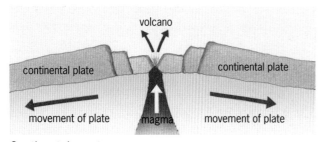

Continental crust

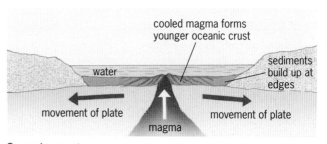

Oceanic crust

If the plates are pushed apart by magma forcing its way through the crust, then in other places they must be pushed together – they collide. If the plates under the continental crust collide, **fold mountains** are formed and the **sedimentary rock** between them is crumpled upwards. The Pyrenees (between Spain and France) and the Alps (on the Eastern side of France) were formed in this way.

As a result, newer mountain ranges are formed and replace (over millions of years) those that are worn down by weathering and erosion.

When oceanic plates collide, something quite different happens. The heavier oceanic crust is forced down below the continental crust, forming mountain ranges as the sedimentary rock at the surface is 'scraped off'. A deep trench forms in the ocean floor where the two plates meet.

As the oceanic crust is pushed down it melts, becoming less dense. It then floats up to the crust where it breaks through weakened areas to form chains of volcanic islands such as those around Japan.

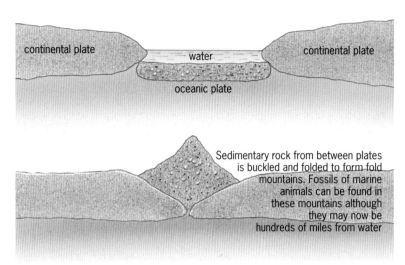

Sedimentary rock from between plates is buckled and folded to form fold mountains. Fossils of marine animals can be found in these mountains although they may now be hundreds of miles from water

Continental plates squeezing together

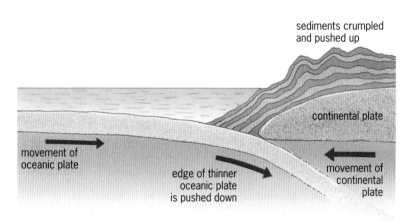

Oceanic and continental plates colliding

★ THINGS TO DO

1 Europe and North America are moving apart at about 3 cm every year. In 1642, Christopher Columbus made the first crossing of the Atlantic Ocean. Use a calculator to work out how much wider the Atlantic Ocean is now compared with when Columbus crossed it.

2 Trace the shape of the continents from a map of the world. Use your tracing to cut out some card shapes of each continent. Do they fit neatly together into one large continent, supporting the idea of Pangaea? If not, can you explain why?

3 Draw the West Coast of Africa and the East Coast of South America. Make a list of reasons why it seems likely that they were once joined together.

4 By agreement between countries, minerals which exist under the Antarctic are not extracted from the Earth's crust. One of the minerals is coal. Coal is formed from the partially decayed remains of plants. How could coal now be found under an area with such an inhospitable climate?

Plate tectonics

The illustration shows the position of the boundaries between tectonic plates across the world.

The position of the plate boundaries, and the direction in which they are moving, help us to explain some of the events which take place on and beneath the surface.

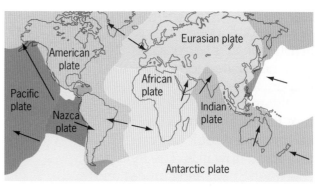

The world's plate boundaries. The arrows show the directions in which the plates are moving

Earthquake

The area around Los Angeles suffers regular earth tremors – slight shakes which cause little damage. Every so often, however, the area suffers from a major earthquake. In 1994, people were killed and buildings and roads destroyed, by one of the biggest earthquakes they had experienced.

This area is prone to earthquakes because it lies on the San Andreas fault – the boundary between the Pacific plate and the American plate. The plates are moving almost parallel to one another, but in opposite directions.

The rough edges grind away as the plates slide across one another but occasionally, particularly rough sections meet. The pressure builds up until suddenly, they slip across one another, causing the ground on each side to shudder violently, causing earthquake damage on the surface.

High in the mountains

Along the western edge of South America you can see a boundary region between two plates that are moving in opposite directions. Continental crust will meet oceanic crust at the boundary between the two plates. When this happens the thinner oceanic plate is pushed down below the thicker continental plate and melts.

Lateral displacement of rows of trees due to plate movement along the San Andreas fault

Sediments scraped from the oceanic plate are pushed up to form fold mountains, such as the Andes. Close to the edge of the continental plate the crust is weakened. Magma forces its way through these areas of weaknesses, forming volcanic regions.

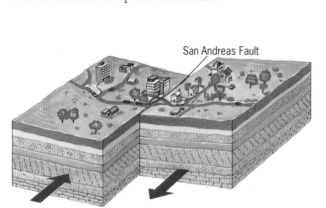

The San Andreas fault

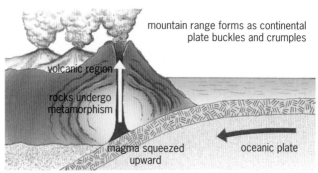

Formation of mountain ranges

The spreading oceans

Although it may be difficult to believe, there are mountain ranges under the sea which are much higher than those on land. They occur where two plates are moving apart. One such region lies in the middle of the Atlantic ocean, as you can see from the world map.

Along the line where the plates meet, lava pours through huge fissures, or cracks. The magma cools quickly in the colder depths of the ocean, forming 'bubbles'. These are quickly burst by the pressure of the lava below, and fresh lava erupts on top of the previously erupted and now hardened layer. Layer upon layer of basalt is formed, forming a high ridge called the Mid-Atlantic Ridge. As more magma emerges, the older rock layers are pushed further and further apart – a process known as **ocean-floor spreading**.

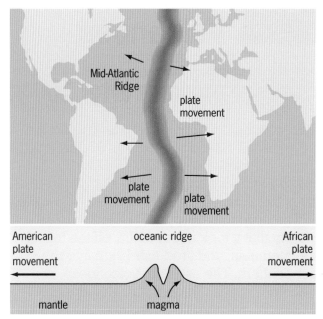

Formation of the Mid-Atlantic Ridge

Magnetic stripes

On each side of ocean ridges the rocks show a clear pattern of **magnetic** stripes. These are caused by iron crystals in the magma which, as it hardens, line up in the direction of the magnetic North and South poles. The iron particles themselves are weakly magnetised, with their North and South poles aligned in the Earth's magnetic field.

At times during the Earth's history, the magnetic North and South poles have suddenly reversed. The reversal of the magnetically aligned particles in the stripes confirms this, but also shows that the basalts on each side of the ridge were intruded into the ridge and became magnetised before being broken in two and moving apart.

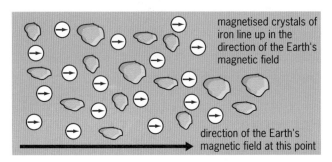

The iron crystals line up with Earth's magnetic field

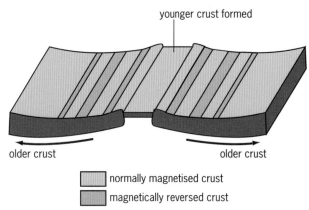

Magnetic stripes in the Earth's crust

★ THINGS TO DO

1 Fossils of sea creatures are found high up on Mount Everest. What does this suggest about how the Himalayan mountain range was formed? Explain your answer.

2 It is estimated that the Pacific Ocean is getting wider by 16 cm each year. How much wider will the Pacific Ocean be in 1000 years' time?

3 What evidence could be gathered to show that the east coast of America and the west coast of Europe were once joined?

Volcanoes

Daily Planet News

Sleeper awakens after 100 years

This was the headline which announced the eruption of Mount St Helens, in Washington State, USA, on 18 May 1980. When it exploded it sent huge amounts of volcanic ash over 1000km and left a crater over 3km wide.

The eruption at Mount St Helens was so violent that it took about 400m off the top of the mountain, and fired ash 19km into the atmosphere. The ash from volcanoes becomes spread through the atmosphere, which can reduce the intensity of the Sun. In 1980 England had a very poor summer!

The size of a volcanic eruption is determined by the pressures below the surface of the crust and the amount of water and gas (such as carbon dioxide) dissolved in the magma. Some eruptions are not as violent as the one at Mount St Helens with the **lava** seeping out of **vents** and **fissures**.

What causes a volcano?

In the crust, rocks are pushed down by the weight of rocks above them when tectonic plates move towards each other. As the rocks are pushed down into the Earth they become molten due to the high temperatures which exist at these lower levels of the crust. These molten rocks are known as magma. Because the magma is a liquid it is less dense than the surrounding solid rock and so begins to rise upwards towards the surface of the crust. This molten rock material, containing dissolved gases and water beneath the Earth's crust, escapes to the surface through any areas of weakness in the crust. These weaknesses can be in the form of cracks (fissures) or holes (vents).

The magma appears at the surface as lava (see illustration). Lava flow can engulf vast areas of land around the volcano. Some volcanoes also produce toxic gases such as sulphur dioxide and hydrogen sulphide which will lead to a loss of life.

Mount St Helens

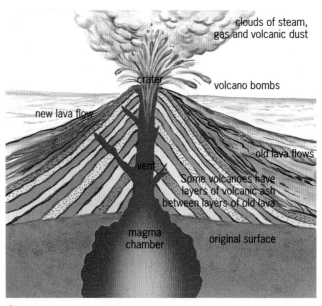

clouds of steam, gas and volcanic dust

crater

volcano bombs

new lava flow

old lava flows

vent

Some volcanoes have layers of volcanic ash between layers of old lava

magma chamber

original surface

Cross section through a volcano

Hot rocks

Rocks which are formed from solidified molten rock are called **igneous rocks**.

Some igneous rocks are formed when magma bursts through the Earth's crust. They are called **extrusive igneous rocks**. They cool rapidly, forming rocks such as basalts. Others are formed below the surface – underneath other rocks. As they cool, crystals form. Because they cool rapidly, the crystals in these rocks are so small they often cannot be seen. The rock may look smooth and glossy.

The Giant's Causeway – extrusive igneous rock

The Scottish Highlands – intrusive igneous rock

At other times, molten rock may be forced up into the Earth's crust but does not break through the surface. These are **intrusive igneous rocks**. Parts of the Scottish Highlands are composed largely of rocks formed in this way. As the surface is eroded, the underlying granite is exposed.

Because these rocks cool more slowly, the crystals are much bigger. In this sample of granite, the crystals can be clearly seen and are completely randomly arranged – there is no pattern to them. You can also see that there are several different minerals present in the rock.

★ THINGS TO DO

1 Use a hand lens to look closely at some granite and basalt. Draw a detailed sketch of what you see. Make a note of any similarities and differences you can find.

2 Under the supervision of your teacher, melt some salol in a test tube by placing the test tube in a beaker of hot water. When the salol has melted you can pour a small amount onto a microscope slide and watch the crystals form by looking through the microscope.

Draw pictures showing the crystals forming.

Try to find out how the size of the crystals depends on how quickly the salol is cooled. You will need to think about how you can alter the cooling time.

3 These photographs all show igneous rocks.

You may be able to observe samples of these rocks with a hand lens if your school has them.

Make a note of what you see.

Can you tell which are intrusive rocks and which are extrusive? What features of the rocks give clues to how they were formed?

Obsidian

Pegmatite

Granite

Basalt

4 For a sample of six igneous rocks make a key which could be used to identify them. Test your key on a friend and modify it if necessary.

The atmosphere forms

About 4500 million years ago the planets were formed. Each planet had a thick layer of gases, mainly hydrogen and helium, surrounding its core – known as the **primary atmosphere**. Intense **solar activity** caused these lighter gases to be removed from the primary atmosphere of **planets** near to the Sun.

At this time the molten Earth had cooled sufficiently for a thin crust to form. Volcanoes on the surface produced huge quantities of gases such as ammonia, nitrogen, methane, carbon monoxide, carbon dioxide and sulphur dioxide. These formed a **secondary atmosphere** around the Earth.

Most of the atmosphere at the time was ammonia and methane gases. The temperature was so high that these gases reacted with any oxygen present to form carbon dioxide, water vapour and nitrogen.

$$ammonia + oxygen \rightarrow nitrogen + water$$
$$4NH_3(g) + 3O_2(g) \rightarrow 2N_2(g) + 6H_2O(g)$$

$$methane + oxygen \rightarrow carbon\ dioxide + water$$
$$CH_4(g) + 2O_2(g) \rightarrow CO_2(g) + 2H_2O(g)$$

About 3800 million years ago, when the Earth had cooled below 100 °C, the water vapour condensed. This formed the first oceans, lakes and seas on the surface of the rapidly cooling Earth. Early forms of life developed in the water at depths which prevented potentially harmful radiation from the Sun affecting them. About 3000 million years ago simple **algae**-like plants appeared along with the first forms of **bacteria**. These algae used the light from the Sun, via **photosynthesis**, to produce their own food. The process also released oxygen into the atmosphere.

The process of photosynthesis can be described by the following equation:

$$carbon\ dioxide + water \xrightarrow[chlorophyll]{sunlight} glucose + oxygen$$
$$6CO_2(g) + 6H_2O(l) \longrightarrow C_6H_{12}O_6(aq) + 6O_2(g)$$

At this time the **ultra violet radiation** from the Sun was so intense it was harmful to living things. But it also served another purpose – it helped to change some of the oxygen into ozone gas.

The atmosphere around the Earth formed over many millions of years

The hostile Earth during the formation of its atmosphere

$$oxygen\ atom + oxygen\ molecule \rightarrow ozone$$
$$O(g) + O_2(g) \rightarrow O_3(g)$$

The ozone gas formed a layer around the Earth which filtered out most of the harmful radiation.

Over many millions of years the intensity of the ultra violet radiation was reduced as the amount of ozone in the atmosphere increased, making it less harmful to living things.

About 400 million years ago the first land plants appeared on the Earth and so the amount of oxygen and ozone began to increase.

Not only did the amount of oxygen begin to increase, but the amount of carbon dioxide began to decrease as it was absorbed by the plants. Combined with the absorption of carbon dioxide by chemical reactions which produced carbonate compounds, this reduced the level of carbon dioxide to levels which could be tolerated by a wider variety of living things.

Some of the carbon dioxide absorbed by plants became 'locked' in the plants when they died. Over millions of years, the partially decayed remains were transformed into fossil fuels – substances which contain high amounts of carbon. When we burn fossil fuels, the same carbon is released back into the atmosphere as carbon dioxide gas. The carbon dioxide locked up in carbonate compounds is released as they are heated deep within the Earth's crust and is returned to the atmosphere by erupting volcanoes.

About 350 million years ago simple animals began to develop which did not rely on sunlight for their energy. Instead they adapted to make use of the oxygen gas in the 'new' atmosphere. The oxygen was used to release energy from food in a process known as **respiration**.

respiration

glucose + oxygen ⟶ carbon + water + energy
dioxide

$$C_6H_{12}O_6(aq) + 6O_2(g) \longrightarrow 6CO_2(g) + 6H_2O(l) + energy$$

Some harmful radiations are absorbed in the ozone layer. Less harmful radiation passes through.

Earth

The ozone layer forms at 25–50 km above the surface of the Earth.

The ozone layer reduces the amount of harmful radiation reaching the Earth

Mariopteris, a fossil plant from the Carboniferous Period

It is this process which keeps us alive today. It is the process by which our bodies release energy from the food we eat.

In the last 200 million years a balance has been achieved between what is needed by living things to survive, and the waste products which they produce. If living things are to survive on the planet this balance must be maintained.

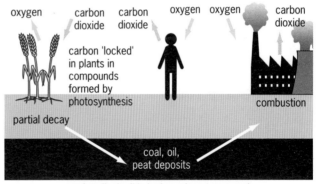

carbon 'locked' in hydrocarbon compounds

The absorption and release of carbon dioxide

★ THINGS TO DO

1 The air around the Earth contains the following gases: nitrogen 78%; oxygen 21%; argon 0.9%; other gases, such as ozone, carbon dioxide and water vapour 0.1%.

Draw a bar chart using this information.

2 The average surface temperature on the surface of Mars is −23 °C. The Martian atmosphere consists of: carbon dioxide 95%; nitrogen 3%; oxygen slightly less than 1%; others, largely water vapour less than 2%.

What would you say to someone who said, 'We should grow plants on Mars and its atmosphere would eventually become like our own. We could then live there.'

3 There is a real danger that the things humans do may change the atmosphere and damage peoples' health. Make a list of the ways in which we pollute the atmosphere. Alongside each one suggest how we could reduce the pollution.

New land from old

The Bay of Bengal

The coastal area of Bangladesh on the edge of the Bay of Bengal has formed slowly over many years, formed by **silt** (small particles of rock) carried down by the River Ganges. As the river widens, the water flows more slowly. Solid materials are deposited on the river bed, building up to form small flat islands. Land formed in this way is called a **delta**.

The map shows the area round the Bay of Bengal as it is today.

Sedimentary rock

Eventually the **sediments** deposited in the Bay of Bengal will form sedimentary rock. Sedimentary rock covers approximately 75% of the continents. It is formed by layers of sediments which are deposited on top of one another over thousands of years. Younger sedimentary rock is therefore always formed on the top of older rock. Layers of sediments gradually build up, pressing down on the oldest layers underneath.

Sandstone

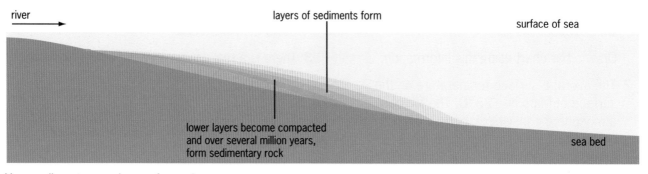

How sedimentary rocks are formed

The water in these layers is squeezed out and the sediments become cemented together by salts which crystallize out of the water. After several million years the sediment layers become bound together, forming new rock.

You can often see separate grains (small particles) in sedimentary rocks. There is a large variation in their hardness and grain size. In some sedimentary rocks the separate layers can be seen clearly. Sedimentary rocks often contain fossils (see photograph opposite) and due to the layering of this type of rock it is often possible to date the fossil according to the layer in which it is found. The fossils found are those of plants and animals that lived at the same time the sediment was laid down.

Not all sedimentary rocks are made from older rocks which have been weathered. Some, such as coal, are the remains of trees and other plants which have taken millions of years to decay. Others, such as limestone, were formed from the crushed remains, generally the shells, of sea animals.

The white cliffs of Dover have always been a welcoming sight to travellers returning to Great Britain. These cliffs are chalk – a sedimentary rock formed from the crushed shells of sea creatures which lived in the warm seas millions of

Fossils are often found in limestone

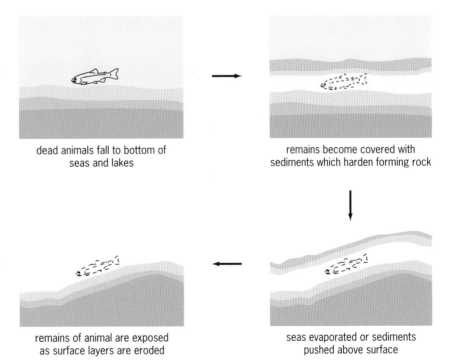

dead animals fall to bottom of seas and lakes

remains become covered with sediments which harden forming rock

remains of animal are exposed as surface layers are eroded

seas evaporated or sediments pushed above surface

How a fossil is formed

years ago. It is a soft rock which weathers rapidly.

In places the newly formed chalk deposits were covered with other rocks and compressed. The sediments became compacted, forming limestone.

Although limestone and chalk are the same compound – calcium carbonate – they have quite different properties due to the way they were formed. Limestone, for example, is much harder than chalk.

The white cliffs of Dover

Limestone cliffs in Dorset

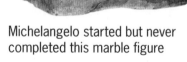

Michelangelo started but never completed this marble figure

Geological time

Fossils are found in sedimentary rocks such as limestone. Sedimentary rocks are layered with the oldest layers (strata) found deeper underground than the younger rock. The fossils in the layers close to the surface are therefore younger than those below them.

Fossils are the remains or impressions of living organisms made when animals or plants die in soft silt. As they decay they leave their impressions in the silt. In some cases the organism decays and dissolves in the silt, leaving a space. Minerals may seep into the

space and harden, taking up the shape of the dissolved organism. When the silt turns into rock a fossil mould is preserved.

By examining the fossils found in the different rock strata, Geologists have been able to divide time into three **eras** according to the type of fossil found (see diagram on page 67): the Cainozoic era, the Mesozoic era and the Palaeozoic era. The diagram shows the way in which the eras have been divided into **periods** and the periods can be further subdivided into **epochs**. This is called **geological time**.

ERAS	PERIODS	EPOCHS	Present	SOME FOSSILS
CAINOZOIC The Earth's climate became much colder, resulting in several Ice Ages	QUATERNARY	Holocene 0.01 Pleistocene 2		human skull mammoth (tooth)
The age of the mammals as well as insects and flowering plants Opening of the North Sea	TERTIARY	Pliocene 7 Miocene 26 Oligocene 38 Eocene 54 Paleocene 65		snail bivalve shellfish
MESOZOIC The age of the dinosaurs	CRETACEOUS	136		ammonite sea urchin lampshell
The great continent of Panagea broke up, forming most of the continents as we know them	JURASSIC	190		lamp-shell ammonite coral sea urchin
The Earth's climate was generally warm and pleasant	TRIASSIC	225		bivalve shellfish fish (tooth)
PALAEOZOIC	PERMIAN	280		bivalve shellfish fish (tail) lamp-shell
Initially, most life was in the sea. Plants appeared on the land in the Silurian era, followed after a few million years by the first amphibians. Towards the end of this period the first reptiles appeared on land	CARBONIFEROUS	355		tree root coral amphibian (skull)
	DEVONIAN	395		lampshell fish
	SILURIAN	440		coral trilobite lampshell graptolite
	ORDOVICIAN	500		graptolite lampshell trilobite
	CAMBRIAN	570		lampshell trilobite trilobite

Geological time (figures refer to millions of years before the present)

★ THINGS TO DO

1 Compare several types of sedimentary rock using a hand lens. Make a list of any similarities and differences which you notice.

2 Although rocks might change, the fossils inside them do not. The fossils in two areas over 100 km apart were studied. The same fossils were found in different types of rock, but fossils of the same type must have lived on Earth at the same time. The picture on the left shows what was found.

a) Put the rocks in order of their age, with the youngest at the top.

b) Suggest why the same fossils could be found in different types of rock.

c) Draw what you think the rocks could look like between these two places.

3 The dinosaurs are perhaps the best known reptiles. Some people suggest that they died out very quickly. What could have caused the extinction of the dinosaurs? What evidence could be found from fossils or rocks to support your idea?

area a
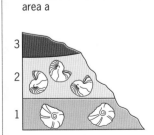

area b
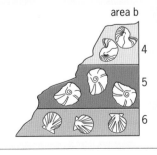

Metamorphic rocks

Some limestones have been pushed much deeper into the Earth than others. As a result they experience high temperatures *and* high pressures. The two factors combine to change the limestone into marble – another form of calcium carbonate. Marble is much, much harder than limestone. Marble is an example of a **metamorphic rock** – one which has been formed by the action of high temperature and pressure on another type of rock.

Metamorphic rocks are also formed when magma forces its way into the crust. The magma heats the surrounding rocks, changing the structure of the particles within them. In general, metamorphic rocks are harder and more resistant to weathering than sedimentary rocks.

Why do the rocks have different properties?

Although rocks do not melt as they change into metamorphic rocks they become very soft, almost putty-like. Crystals inside the rocks are squeezed into parallel layers. It is these changes which account for the different properties of the parent rock and the metamorphic rock formed from it.

Quartzite is formed when grains of sand undergo a series of changes caused by high pressure and temperature (see illustration). The metamorphic rock which is formed is quite different to the sand from which it was made.

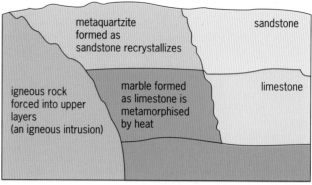

The formation of metamorphic rocks

Quartzite

The formation of quartzite from sand

sand — sand grains loosely and randomly arranged

sedimentary sandstone — under pressure, sand grains are compacted (squeezed together) randomly

metamorphic quartzite — under pressure and heat the grains reform and align in parallel bands

In several different forms

Under low temperatures and high pressure, mudstones and shales form slate, a rock which was commonly used for roofing. Slate shows clear evidence that the particles have been compressed and realigned into layers.

When slate is formed the minerals' structure changes. Clay particles form flakes of crystals called mica. Mica crystals are very thin and turn so that their flat face is at right angles to the applied pressure, so clear layers are seen in the newly formed slate.

Splitting slate

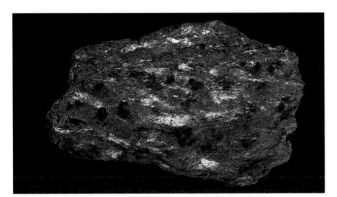

Schist

This gneiss on Iona is over 2700 million years old

At higher temperatures and high pressure the mudstones and shales do not form slate, but form schist. In schists the mica flakes are much bigger. They also grow in parallel bands.

At still higher temperatures the mudstones and shales form a different type of rock again

– gneiss. Gneiss looks rather like granite, but, unlike the granites, the layers within the gneiss are quite clearly banded. Crystals of mica, feldspar and quartz may also be present, caused by changes in the minerals present in the original mudstones or shales.

★ THINGS TO DO

Use the data in the table opposite to answer the questions which follow.

Rock	Type	Force required to crush it (relative)	Impact strength (relative)
Limestone	Sedimentary	2.5	0.5
Flint	Igneous	9	23
Schist	Metamorphic	14	15
Sandstone	Sedimentary	8.5	16
Quartz	Metamorphic	27.5	21

1 Which type of rock seems to be the strongest? Which data did you use to come to this conclusion?

2 Draw an apparatus which could be used to obtain the data for the **impact strength** figures.

3 Why do you think there is such a large difference in the strength of sedimentary and metamorphic rocks?

4 Give a use, past or present, for flint, sandstone and limestone. For each type of rock, state which property allows it to be used for the use you have given.

A cycle of change

The movements of rocks are part of a huge cycle of events which is taking place all the time. This cycle of events is called the **rock cycle**.

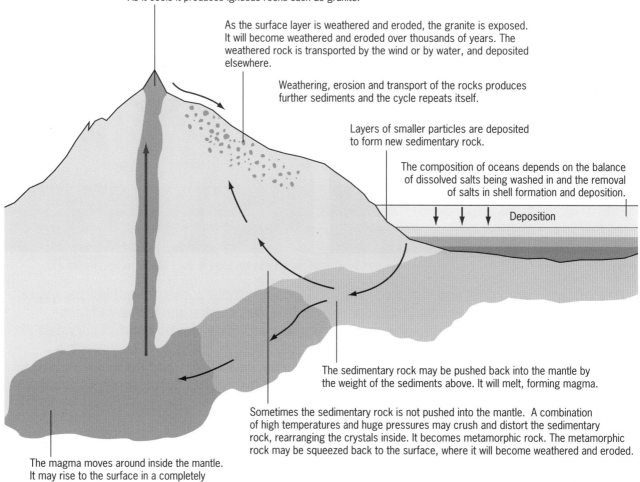

Molten rock forces its way into the upper layers of the Earth's crust. As it cools it produces igneous rocks such as granite.

As the surface layer is weathered and eroded, the granite is exposed. It will become weathered and eroded over thousands of years. The weathered rock is transported by the wind or by water, and deposited elsewhere.

Weathering, erosion and transport of the rocks produces further sediments and the cycle repeats itself.

Layers of smaller particles are deposited to form new sedimentary rock.

The composition of oceans depends on the balance of dissolved salts being washed in and the removal of salts in shell formation and deposition.

Deposition

The sedimentary rock may be pushed back into the mantle by the weight of the sediments above. It will melt, forming magma.

Sometimes the sedimentary rock is not pushed into the mantle. A combination of high temperatures and huge pressures may crush and distort the sedimentary rock, rearranging the crystals inside. It becomes metamorphic rock. The metamorphic rock may be squeezed back to the surface, where it will become weathered and eroded.

The magma moves around inside the mantle. It may rise to the surface in a completely different place.

Above The rock cycle

Right Scree slope formed by physical weathering

★ THINGS TO DO

1 Why is the picture opposite called the rock *cycle*?

2 Draw a flow chart showing the changes which occur at each part of the cycle. The first one should be:

Igneous rock (molten)
flows from the mouth of →
an active volcano

Try to use these words in your flow chart:

melted	solidified
melting point	weathering
erosion	transport
deposition	sedimentary rock
igneous rock	metamorphic rock

3 Make a crossword using the words from question **2**. Write the clues for the words. Let one of your friends try your crossword.

4 Simple **keys** can be used to sort rocks according to whether they are igneous, sedimentary or metamorphic. The key below can be used.

a) Use the key to identify some samples of different rocks which your teacher will provide.
b) Make a table showing the key features of igneous, sedimentary and metamorphic rocks.

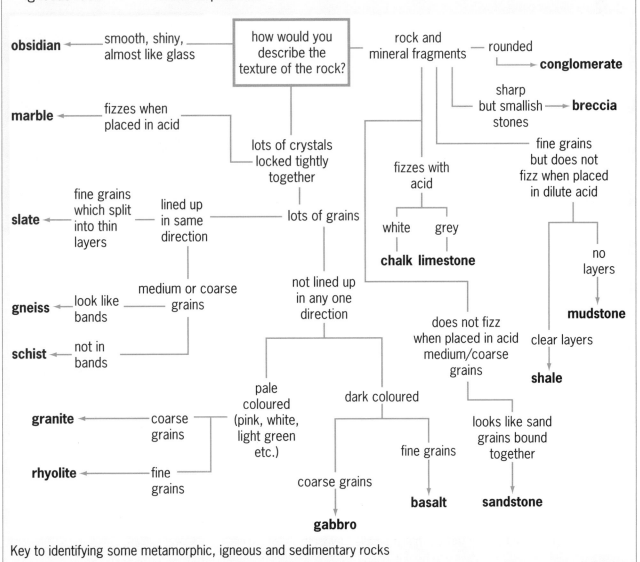

Key to identifying some metamorphic, igneous and sedimentary rocks

Sorting rocks

The rocks in the Earth's crust contain chemical compounds which provide many of the raw materials from which thousands of useful products are made.

The useful chemicals in the rocks are called minerals. Some minerals contain metal compounds from which we obtain most of the metals we use every day.

Identifying minerals is a big problem. Their appearance is not really a reliable way of identifying them because so many of them look alike. Other properties must also be used.

Using a key

You may have used a key in biology (see *GCSE Science Double Award Biology*, Topics 3.4, 4.1) to identify plants and animals. Any key is a series of steps which, if followed, should allow you to identify different things, including different types of rock. At each step you follow the most appropriate route through the key by, for example, observing the texture or colour of a specimen.

The photographs on page 73 show students testing the properties of some minerals.

They used the equation

$$\text{density} = \frac{\text{mass}}{\text{volume}}$$

to find the **density** of each mineral.

There are lots of other tests you could do.

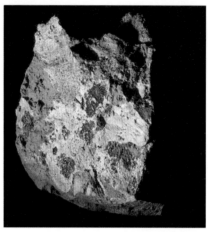

Iron is obtained from this ore called haematite

Copper is obtained from an ore called malachite

Sulphur is obtained from this mineral called pyrites

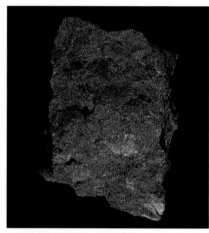

Mercury is obtained from cinnabar

Plaster of Paris is made from a mineral called gypsum

Corundum is the second hardest mineral known. This ruby is a natural form of corundum

We tested each mineral to find out whether they reacted with dilute hydrochloric acid.

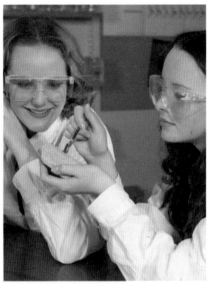

We tested each mineral and put them in order of hardness

We made notes about what the mineral looked like and what it felt like. We used a magnifying lens to help us see the grains and crystals

★ THINGS TO DO

1 Have a class discussion about the information you would need to design a good key which would help other people identify the following minerals: iron pyrites, haematite, magnetite, graphite, galena, malachite, sulphur, calcite, dolomite, feldspar, quartz.

Share out the tests which your class thinks would be useful.

When everyone has done their tests design a table which could be used to record all the information your class found out.

When you have finished, make a key which could be used to identify the minerals you tested. Test your key on one of your friends.

2 Design a mineral database which could be used. Include an information leaflet describing how to use your database.

3 The water which you drink has probably flowed through minerals in the Earth's crust. As it does so some of the minerals dissolve in the water. 'Mineral water' is water containing dissolved minerals.

a) What does the label mean when it says 'volcanic rocks constantly filter Volvic, naturally purifying the water'?

b) Which rocks does the water filter through?

c) What does 'Total Dry Residue at 180°C = 109mg/l' mean?

d) Why does the label say 'Suitable for a low sodium diet.'?

e) Look at some bottles of mineral water in your local supermarket. Make a note of which minerals they contain.

In the heart of France, beneath the ancient volcanoes of the Auvergne National Park, layers of volcanic rocks constantly filter Volvic, naturally purifying the water and imparting their precious minerals to give Volvic its unique composition and delicate taste.
So by simply drinking Volvic, you will experience the refreshing legacy of Nature's goodness.
Volvic Natural Mineral Water : naturally for everyone !

volvic
OFFICIAL ANALYSIS · 20.04.1989 (mg/litre)

Calcium	9.9	Aluminium	< 0.01
Magnesium	6.1	Chlorides	8.4
Sodium	9.4	Nitrates	6.3
Potassium	5.7	Sulphates	6.9
Iron	< 0.01	Bicarbonates	65.3

Total Dry Residue at 180°C = 109 mg/l. pH7
Suitable for a low sodium diet.
Daily adult water requirement : 2.5 litre. 1 of which should be provided through food and 1.5 by drinking water.
For perfect condition, drink one full 1.5 litre bottle of Volvic per day. Enjoy !
Store in a cool, dry and clean place away from light. Best before end 1997. Bottle exclusively designed for the use of Volvic Natural Mineral Water. Do not refill.

PREMIER WATERS Ltd,
4, Hillgate Place, London SW12 9 ER.
A Company of the DANONE GROUP

Fossil fuels

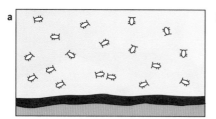

Millions of years ago the seas were warm and teemed with marine animals and plants. When they died they formed a thick layer on the sea bed.

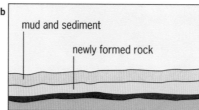

Over millions of years, layer upon layer of sediment covered the remains of the animals and plants. The weight of the sediments subjected the layers of animal and plant remains to huge pressures. Because there was little oxygen present the remains of the animals and plants only partly decayed. At high temperature and pressure they formed gas and oil.

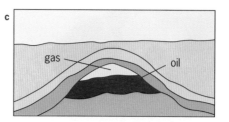

The sediment slowly changed to sedimentary rock. The layers of rock were folded by movements in the Earth's crust. The oil and gas moved through some layers until they met harder, impervious rock through which they could not pass. They became trapped under the rock.

The formation of oil and gas

Fossil fuels – coal, oil and gas – are some of the most important substances found in the Earth's crust. Until the introduction of nuclear power, fossil fuels provided all of the energy needed for our homes, for transport, for industry, and to make electricity. Even today, about half of the electricity made in generating stations is made from fossil fuels.

All fossil fuels are formed from the partly decayed remains of plants and animals which lived on land and in the seas millions of years ago. Oil, for example, was formed from the remains of marine life.

By drilling holes through the surface rock into the pockets of oil and gas, they are released and brought to the surface. Eventually a pocket of oil and gas will be emptied. Exploration teams continually search for undiscovered reserves of oil and gas so that the world's needs for fossil fuels can continue to be met.

A raw material

Naturally occurring oil is called **crude oil** – quite different from the oils you see around you. It is a very important raw material from which we obtain many of the things which we use every day such as those shown in the photograph.

In its natural state crude oil is of little use. It is a mixture of many different chemical compounds which are much more useful when they are separated from one another. Most of these compounds are **hydrocarbons** (compounds formed when atoms of the elements hydrogen and carbon join together). Each carbon atom can form four bonds with other atoms (including other carbon atoms). Carbon forms long chains with other atoms, particularly those of hydrogen, joined to the sides.

Products made from crude oil

There is a huge range of hydrocarbon compounds – called **organic compounds** – based on the long carbon chain. Some typical examples of these compounds are shown in the illustration opposite.

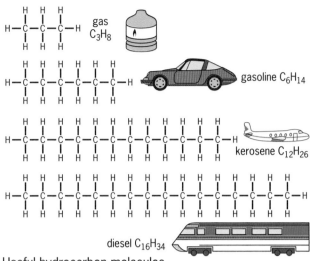

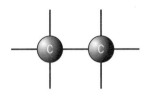

You can see that the structure of each molecule is similar to that of the others, with two hydrogen atoms attached to each carbon atom in the middle of the chain, and three hydrogen atoms joined to the carbon atom at each end of the chain.

These hydrocarbon molecules have a wide range of properties. To some extent, the properties depend on the number of carbon atoms in the chain.

Useful hydrocarbon molecules

Because they are more difficult to ignite, compounds with larger hydrocarbon molecules are of little use as fuels. They do, however, provide many more useful substances when they are 'cracked' into smaller molecules (see page 85).

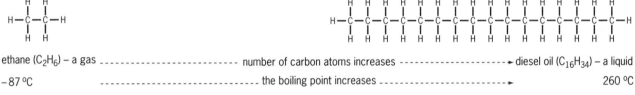

ethane (C_2H_6) – a gas	number of carbon atoms increases	diesel oil ($C_{16}H_{34}$) – a liquid
–87 °C	the boiling point increases	260 °C
vaporises easily	become more difficult to vaporise	vaporises slowly
flows easily	flow less and less easily	thicker than gas
easily lit	become increasingly difficult to ignite	difficult to ignite

The properties of hydrocarbons depend on the number of carbon atoms they contain

---

★ **THINGS TO DO**

1 Draw a flow chart showing the stages in the formation of crude oil.

2 When fossil fuels burn, the carbon and hydrogen atoms react with oxygen in the surroundings to form carbon dioxide and water.
   Design a leaflet warning householders of the dangers of burning natural gas (gas fires) in unventilated rooms.

3 Copy and complete the table opposite in your notebook. Add as many hydrocarbons as you can.

Hydrocarbon	Molecular formula	Boiling point	Ease with which it flows
Ethane	C_2H_6 H—C—C—H		

4 Someone said: 'When fossil fuels are burned they release carbon dioxide gas. The carbon atoms go back into the air, which is where they came from in the first place.' Explain what you think they meant.

Very useful

To make crude oil more useful it must first be separated into groups of substances which have similar properties. These groups of substances are called **fractions**. Each fraction is a part of the complicated mixture of compounds we call crude oil. A typical barrel of crude oil from the North Sea would contain the fractions shown in the illustration. Oil from other parts of the world contains different amounts of each fraction.

Because they have different boiling points, each fraction (group of compounds) in the mixture can be separated by evaporating the oil and **condensing** the vapour at different temperatures. This process, known as refining, is carried out at oil refineries.

Imagine, for example, you have a mixture of just two compounds with boiling points of 50°C and 80°C. The two compounds in the mixture can be separated from one another because they have different boiling points. A simplified view of how this happens is shown in the illustration. Separating the fractions of a liquid mixture in this way is called **fractional distillation**.

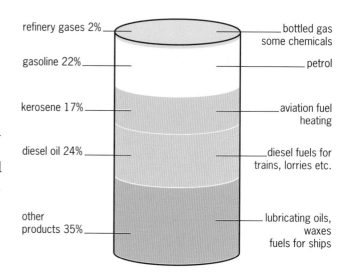

The fractions of a barrel of crude oil from the North Sea and their uses

Fractional distillation is also used to separate fractions in the much more complex crude oil mixture. Each fraction contains several compounds with similar properties and whose boiling points are close to each other.

To separate the different fractions on an industrial scale a fractionating column is used. Each fraction is obtained by collecting hydrocarbon molecules which have a boiling point in a given range of temperatures.

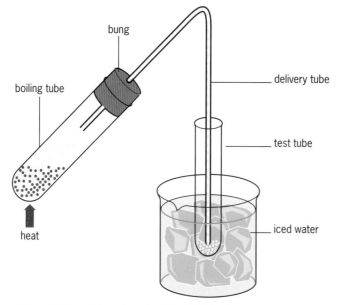

If the temperature of the mixture is raised to 50°C, the yellow fraction boils and turns into a vapour. The brown fraction remains in the liquid state as it does not reach its boiling point of 80°C. The vapour from the yellow fraction is cooled and condenses back into a liquid

An industrial fractionating column

For example, the fraction we know as petrol contains molecules which have boiling points between 30–110°C. The molecules in this fraction contain 5–10 carbon atoms. These smaller molecules with lower boiling points condense higher up the tower. The bigger hydrocarbon molecules which have the higher boiling points condense in the lower half of the tower.

The liquids which condense at different levels are collected on trays. In this way the crude oil is separated into different fractions. These fractions usually contain a number of different hydrocarbons. The individual hydrocarbons can then be separated by refining the fraction by further distillation. We can duplicate this process in the laboratory by using the apparatus shown in the illustration.

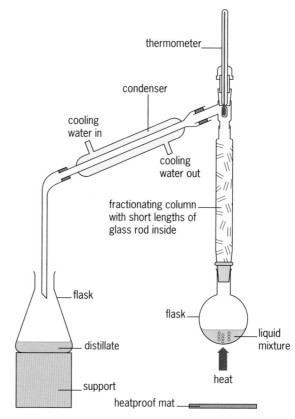

refinery gas
used as fuel

fractionating tower

40°C
gasoline
used as fuel in cars (petrol)

110°C
naphtha
used to make chemicals

180°C
kerosene
used as a fuel in jet engines

260°C
diesel oil or gas oil
used as a fuel in diesel engines

350°C
fuel oil
used as a fuel for ships

400°C
lubricating oil

crude oil

heater

over 400°C
residue
used to make bitumen for surfacing roads

Fractional distillation of crude oil in a refinery

thermometer

condenser

cooling water in

cooling water out

fractionating column with short lengths of glass rod inside

flask

flask

distillate

liquid mixture

support

heat

heatproof mat

Apparatus used to fractionally distil crude oil in the laboratory

★ THINGS TO DO

1 Petrol is one of the substances in the crude oil mixture. Imagine you are a petrol molecule in crude oil. Write a story describing what happens to you from the time you enter the fractionating column to the time you leave. Try to use words such as boiling point, liquid, vapour, condensing.

2 Make a table showing the main fractions obtained from crude oil, their range of boiling points, and their uses.

3 Paraffin and methylated spirits are used as fuels in simple camping stoves. Plan a test you could do to find out which releases most energy when they are burned. Make sure you consider how to make your tests fair and safe.

If possible, under the supervision of your teacher, carry out your tests and prepare a report for a camping magazine describing the advantages and disadvantages of each fuel as a camping fuel.

2.14 Fuels

The early people of the Earth – such as the cave dwellers – appreciated the need for fuels to keep them warm and to cook their food. At the time the only fuel available was wood. Over time, we have discovered coal, gas and oil can also be used as fuels. Today these fuels are amongst the most important substances on Earth, needed not only to warm our homes and offices, but for transport, industry, and to provide the energy needed to generate electricity.

Transport requires vast amounts of fuel to carry food, goods and people from place to place

Fuels supply the energy needed to warm our homes

Fuels supply over half of the electricity we need

Fuels are substances which release energy. Fossil fuels, for example, release energy when they burn in oxygen. The energy is used in many different ways.

Some of the fractions obtained from crude oil ignite, vaporise and flow easily. These properties make them particularly suitable as fuels for engines in vehicles, aeroplanes and ships where they must flow through pipes to be burned in the engines. (Vapours of volatile flammable materials are more dense than air and pose a fire hazard.)Those fractions which have larger molecules are more difficult to ignite, do not vaporise as easily, and are thicker, so they do not flow as well. They are of little use as fuels.

Fuels affect the air around us

When fuels burn, they use up oxygen from the air. This kind of chemical change is called **combustion**. As they do so they produce waste gases such as carbon dioxide and water vapour, which are released into the air.

RESIDENTS' FURY AT ROAD PROPOSALS

Hundreds of people joined a march yesterday to protest at plans for a motorway which would pass within 100 metres of their homes. Many claim that there is now conclusive medical evidence linking the increasing numbers of people suffering from asthma and other respiratory diseases with the pollution from traffic near their homes.

Exhaust fumes – the products of combustion

Fossil fuels are hydrocarbon compounds. When they burn, the hydrogen and carbon atoms in the fuel combine with the oxygen in the air forming their oxides. The carbon atoms join up with oxygen atoms to form carbon dioxide. The hydrogen atoms join with oxygen to form water (hydrogen oxide):

$$\text{hydrogen} + \text{oxygen} \rightarrow \text{water}$$
(from the hydrocarbon in fuel)

$$\text{carbon} + \text{oxygen} \rightarrow \text{carbon dioxide}$$
(from the hydrocarbon in fuel)

Overall the reaction can be summarised as:

$$\text{hydro-carbon} + \text{oxygen} \rightarrow \text{carbon dioxide} + \text{water} + \text{energy}$$

For example, the reaction for methane (natural gas) is as follows:

$$\text{methane} + \text{oxygen} \rightarrow \text{carbon dioxide} + \text{water} + \text{energy}$$

$$CH_4(g) + 2O_2(g) \rightarrow CO_2(g) + 2H_2O(g)$$

We say that the hydrogen and carbon have been *oxidised* – they have formed oxides. The reaction is an *oxidation reaction*. (See page 16 for more about oxidation.)

Energy in fuels

The energy in fuels originally came from the Sun.

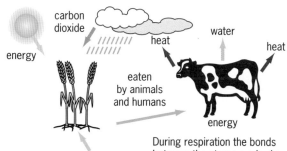

carbon dioxide
energy
heat
water
heat
eaten by animals and humans
energy
water

Green plants absorb energy from the Sun. The energy is used to convert carbon dioxide and water into complex sugar molecules which have lots of carbon, oxygen and hydrogen atoms bonded together. The energy is stored in the bonds which hold the atoms together.

During respiration the bonds between the atoms are broken as the sugars are broken down again into carbon dioxide and water (which is breathed out). The energy which was stored in the bonds (holding the atoms together) is released. Some passes to the surroundings as heat. Some is used to build other complicated molecules which the body needs.

The Sun provides all our energy

★ THINGS TO DO

1 Draw a flow chart showing how energy becomes stored in each of the following fuels: coal, oil, sugar and wood.

2 Some people claim that charcoal briquettes are better than lumpwood charcoal. A group of pupils tested the idea and produced the report shown below.

We decided to find out how much heat was produced by each type of charcoal when it was burned. We weighed the same amount of each type of charcoal and placed it on a steel tray so that it could get plenty of oxygen to burn. Above the tray we placed a metal can containing 500cm³ of water. Each time we did this the water was at 16°C.

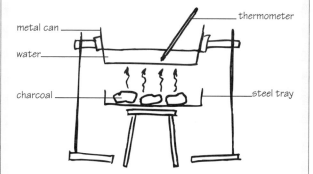

metal can
thermometer
water
charcoal
steel tray

These were our results:

Type of charcoal	Temperature of water before heating (°C)	Temperature of water after heating (°C)
Small lumps	16	36
Pieces of briquette	16	41

a) Make a list of things which the pupils did to make their test fair.

b) What do their results suggest about the different types of charcoal which were tested?

c) Is there anything about the experiment which might cast some doubt on any conclusions which can be drawn from their results?

d) What steps, if any, could be taken to improve their investigation?

The balance in the air

The composition of the atmosphere has stayed the same for over 200 million years. The proportions of the different gases are shown in the pie chart.

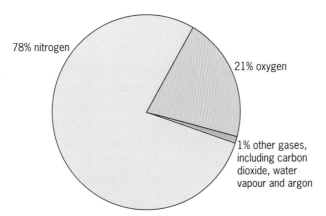

78% nitrogen

21% oxygen

1% other gases, including carbon dioxide, water vapour and argon

The composition of the air

Each time something is burned, and each time an animal breathes in, a small amount of the oxygen in the air is used up. Each time an animal breathes out, or a hydrocarbon fuel is burned, carbon dioxide gas is released, which goes back into the air.

It seems as though the amount of carbon dioxide in the air should increase, and the amount of oxygen in the air should decrease. This has not happened because photosynthesis in plants and respiration in animals have managed to retain a balance in these gases. (See page 62 for more about photosynthesis.)

But just lately

The beginnings of the industrial revolution in the eighteenth century produced a huge demand for fuels to supply the energy needed by machines. Since that time, industry has expanded and the population of the Earth has increased dramatically.

Millions of tonnes of coal, oil and gas have been burned to supply the energy we need. Over the same period, vast amounts of woodland and forest have been chopped down to supply wood. As a result, we now produce more carbon dioxide, but there are

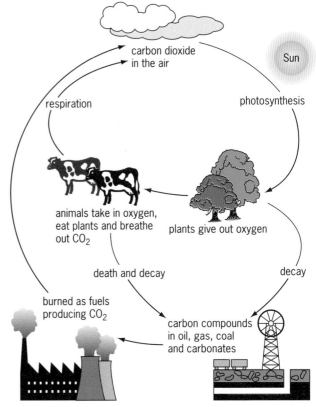

carbon dioxide in the air

Sun

respiration

photosynthesis

animals take in oxygen, eat plants and breathe out CO_2

plants give out oxygen

death and decay

decay

burned as fuels producing CO_2

carbon compounds in oil, gas, coal and carbonates

The carbon – oxygen cycle

O_2

CO_2

COMBUSTION & RESPIRATION

CO_2

O_2

PHOTOSYNTHESIS

The balance of oxygen and carbon dioxide

Pollution in industrial England

Logging in a tropical rain forest

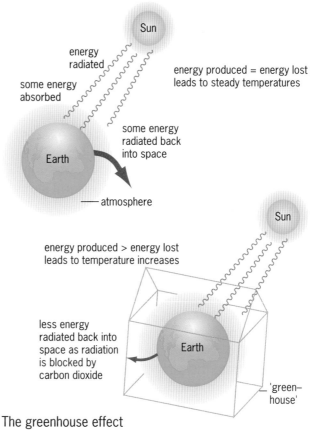

The greenhouse effect

fewer trees to absorb this and replace it with oxygen. Slowly, the amount of carbon dioxide in the atmosphere has started to rise. This increases the '**greenhouse effect**' and may cause global warming – resulting in the average temperature across the surface of the Earth increasing.

The Earth's temperature has remained fairly steady because it radiates (loses) energy as fast as it is produced by the decay of radioactive isotopes in the core. As the amount of carbon dioxide in the atmosphere increases, some of the radiation becomes trapped. Other gases, including methane, contribute to the problem.

In the past 100 years, the temperature across the Earth has been rising slowly.

Although the change in temperature has been very small, it is enough to trigger off much bigger changes in the environment such as:

• climatic changes which would produce drier conditions across the surface, possibly resulting in less food production;

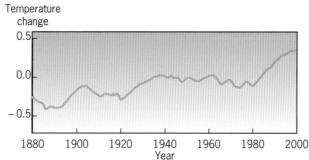

Average world temperature changes in the twentieth century

• a melting of the ice at the North and South poles which would cause the sea levels to rise, flooding vast areas of land.

★ **THINGS TO DO**

1 Produce a newspaper article describing global warming in terms which could be understood by others. You should include: the cause of global warming; why it has only become a problem in the past 200 years; things which we do which increase the problem; what the effects of global warming could be; what can be done to prevent the problem becoming worse.

If possible word process your article. You should include sections of your own research, such as data and articles by other people.

Only a cathedral

In 1986 stone masons began work on Durham cathedral. Their work is expected to take as long as 25 years. Their job is to remove and replace damaged stone.

Some of the damage is due to the effects of weathering. Other damage has been caused by acidic rain (or just **acid rain**). Acids react with some of the compounds in natural stone, dissolving them. This spoils the finer detail in the stonework and can also affect its strength.

What causes acid rain?

When fossil fuels burn they release carbon dioxide gas into the air. Some fossil fuels contain a little sulphur. As the fuel burns, this sulphur is oxidised to sulphur dioxide gas, which is also released into the air. As these gases rise into the atmosphere they dissolve in water vapour, producing acidic solutions which fall as rain.

Damaged stonework

$$\text{carbon dioxide} + \text{water} \rightarrow \text{carbonic acid}$$
$$CO_{2(g)} + H_2O_{(l)} \rightarrow H_2CO_{3(aq)}$$

$$\text{sulphur dioxide} + \text{water} \rightarrow \text{sulphurous acid} \xrightarrow[\text{from air}]{+\text{oxygen}} \text{sulphuric acid}$$
$$SO_{2(g)} + H_2O_{(l)} \rightarrow H_2SO_{3(aq)} \longrightarrow H_2SO_{4(aq)}$$

At high temperatures, such as those inside a furnace or inside a car engine, nitrogen in the air also becomes oxidised, forming nitrogen oxides. These also dissolve in water vapour in the air, producing a weak solution of nitric acid. Acid rain is a cocktail of all three acids.

These gases can also have a direct effect on plants, humans and other animals near where they are formed.

Industries which produce significant amounts of sulphur and nitrogen oxides often build tall chimneys so that the gases are released high in the atmosphere. They then become spread out (and so less concentrated) before reaching ground level. Filters can be fitted which absorb most of the polluting gases, but they are very expensive.

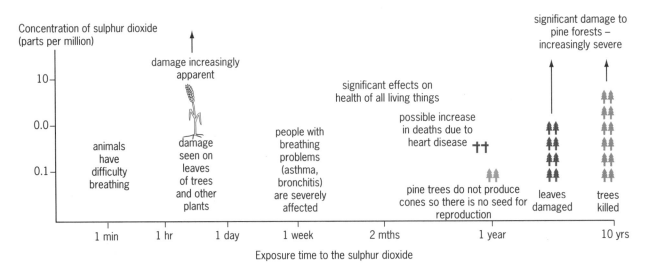

The effects of sulphur dioxide on animals and plants

The dead loch

Little Loch Broom, in the Highlands of Scotland, was popular with tourists. The clean waters were well stocked with trout, and supported many other forms of life. Tall, healthy conifers surrounded the Loch, providing a habitat for many birds and animals. During the 1980s, the trees began to lose their leaves. Other plants turned yellow, then brown, and eventually died. Fish and other animals were found dead on the surface and at the sides of the Loch. Within a few years nothing lived in the water of the Loch. The trees and plants surrounding the Loch were dead.

Tests showed high levels of acidity in the soil and the water. High aluminium levels were also found. The 'death' of the Loch and the surrounding area was considered to be

The devastating effects of acid rain on Loch Broom

due to the effects of acid rain. Steps taken to reduce the level of acidity in the soil and the water, by scattering chemicals which reduce the acidity such as slaked lime (see page 48), are only now beginning to show some success.

★ THINGS TO DO

1 Many lakes especially those in Scandinavia, are treated with chemicals called bases. The base reduces the level of acidity in the lake to an appropriate level. The amount which is used must be carefully controlled. Adding too much can make the waters alkaline. This can cause as much damage as acidity.

Some bases which could be used are: calcium hydroxide, magnesium oxide, magnesium hydroxide and aluminium hydroxide.

Plan tests to find out which of these bases would be best for neutralising an acid lake such as the one described above. You may need to find information about the cost of each substance from a chemical catalogue.

When you have carried out your tests under the supervision of your teacher, prepare a report which could be used by a company specialising in this process.

2 These charts were produced from data supplied by the Department of the Environment and the Forestry Commission.

a) What conclusions can you draw from each chart?

b) Why do you think the amount of sulphur dioxide released into the atmosphere from homes and industry has fallen since 1960?

c) Why do you think less damage to both types of tree was noticed in 1985?

d) Suppose someone said that the charts show that damage to trees is not linked to sulphur dioxide emissions. What would you say to them?

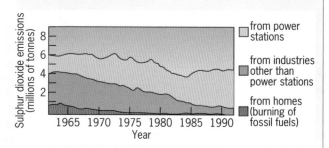

Sulphur dioxide emissions 1960–1990

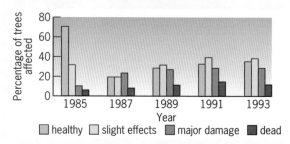

Damage to conifer trees 1985–1993

Alkanes

Most of the hydrocarbons in crude oil belong to a group of compounds called **alkanes**. Alkanes are not generally affected by alkalis, acids or many other substances – they are fairly unreactive. Their most important property is that they burn easily. This makes them very suitable as fuels.

Gaseous alkanes, such as methane (better known as natural gas), burn well in air, forming carbon dioxide and water and releasing energy.

$$\text{methane} + \text{oxygen} \rightarrow \text{carbon dioxide} + \text{water} + \text{energy}$$

$$CH_4(g) + 2O_2(g) \rightarrow CO_2(g) + 2H_2O(g)$$

Butane gas cylinder for use in a domestic heater

Propane and butane are sold as **liquefied petroleum gases (LPG)**. *Propagas* and *Calor Gas* are actually propane and butane. In areas where there is no supply of natural gas, central heating systems can be run on propane gas.

The paraffin used in heaters and lamps is a liquid alkane which also vaporises readily and ignites fairly easily.

The heavier alkanes (those with larger molecules) do not vaporise easily and are not used as fuels. Although they are used to make waxes and lubricating oils there are far more of these fractions in crude oil than we actually need. On the other hand we need more of the lighter fractions, such as bottled gases, petrol and diesel oils than can be obtained from crude oil. To balance demand, some of the heavier fractions are broken, or **cracked**, into smaller, more useful molecules.

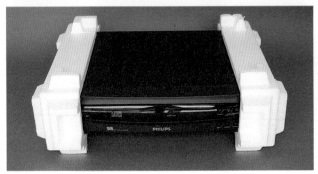

Materials made from alkanes which have been cracked

Cracking up

Some types of crude oil contain large amounts of heavier alkanes which are of little use. They can, however, be broken down (or cracked) into other compounds with smaller molecules which are much more useful. The new compounds can then be used as the raw materials from which we can make thousands of other more useful products.

Ethene gas, for example, is a compound which is produced when the large molecules of Vaseline or medicinal paraffin are broken down. Ethene is then used to make polythene, Terylene, PVC, polystyrene and ethanol (or methylated spirits).

Ethene can be made in the laboratory by cracking medicinal paraffin or Vaseline using the apparatus shown in the illustration.

The broken pottery acts as a **catalyst**. Catalysts are used to speed up chemical reactions.

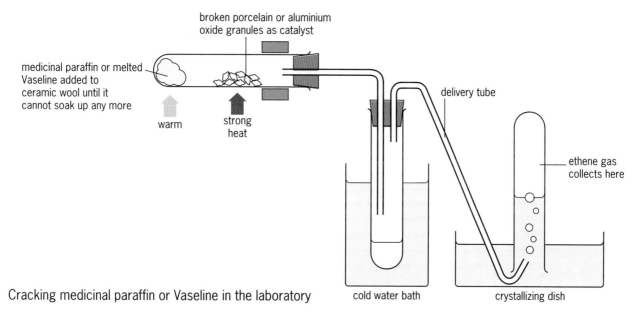

Cracking medicinal paraffin or Vaseline in the laboratory

★ THINGS TO DO

1 There are different types of catalyst which can be used here, such as alumina or sand (silica).

Plan some tests which you could do to find out which catalyst is best. Don't forget to consider any safety precautions you must take. With your teacher's permission carry out your own tests.

Keep a clear record showing what you do and what you found out. Include any evidence which suggests that a chemical reaction has taken place.

2 The table shows the amount of each fraction in oils from the United Kingdom and the Middle East.

North Sea oil		Middle East oil	
gas	2%	refinery gas	2%
gasoline	22%	gasoline	12%
kerosene	17%	kerosene	16%
diesel	24%	diesel	15%
others	35%	residues	55%

Use the information in the charts to help explain why we sell oil to the Middle East and buy oil from there.

Hydrocarbon bonding

Alkanes

The carbon atoms in alkanes are **covalently bonded** (see page 118) to four other atoms by single bonds. Because these molecules possess only single bonds, each attached to another atom, they are said to be **saturated**. No further atoms can be added (which explains why they are so unreactive) although they may be substituted in some reactions.

Some common alkanes are shown below with their properties.

Alkane	Formula	Melting point (°C)	Boiling point (°C)	Physical state at room temperature
Methane	CH_4	−182	−164	gas
Ethane	C_2H_6	−183	−87	gas
Propane	C_3H_8	−190	−42	gas
Butane	C_4H_{10}	−138	0	gas

methane

Methane molecule showing (saturated) single covalent bonds

The illustrations on this page show the actual arrangement of the atoms in methane, ethane, propane and butane.

Alkane compounds all have a similar structure and similar name endings (they all end in -ane). They also react with other chemicals in a similar way and can be represented by a general formula. The general formula for alkanes is $C_nH_{(2n+2)}$ where n is the number of carbon atoms present.

A group of compounds with the above factors in common is called a **homologous series**. As you move through a series such as this the physical properties of the compounds gradually change. For example, boiling points of the alkanes shown in the table gradually increase as the number of carbon atoms increases. This is due to an increase in the **intermolecular forces** (**van der Waals' forces**) (see page 120) as the size of the molecule increases.

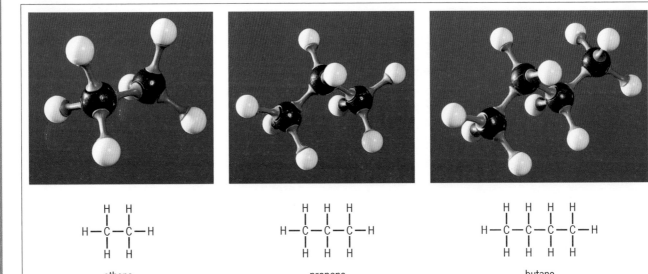

ethane

propane

butane

The molecules look like the models in the photographs

Alkenes

Alkenes are another group of hydrocarbons. Their general formula is C_nH_{2n} where n is the number of carbon atoms.

The simplest examples are ethene (C_2H_4), propene (C_3H_6) and butene (C_4H_8).

The alkenes, especially ethene, are very important industrial chemicals. They are used extensively in the plastics industry and in the production of alcohols such as ethanol and propanol.

The table below shows some physical properties of the first three members of the alkene family.

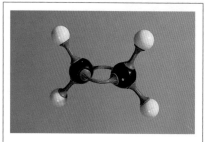
ethene

Alkene	Formula	Melting point (°C)	Boiling point (°C)	Physical state at room temperature
Ethene	C_2H_4	−169	−104	gas
Propene	C_3H_6	−185	−47	gas
Butene	C_4H_8	−184	−6	gas

Once again you can see a connection between the boiling points, and the number of carbon atoms in the molecule.

The alkenes are more reactive than the alkanes because they each contain a **double covalent bond** between the carbon atoms, as shown in the illustration. Molecules which possess a double covalent bond of this kind are said to be **unsaturated** because it is possible to break this double bond and add extra atoms to the molecule.

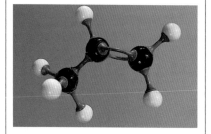
propene

Very few alkenes are found in nature. Most of the alkenes that we need are obtained by breaking up larger less useful alkane molecules. This is usually done by catalytic cracking. In this process the alkane molecules to be cracked (split up) are passed over a mixture of aluminium and chromium oxides heated to about 500°C.

$$\begin{array}{ccccc} \text{dodecane} & \rightarrow & \text{decane} & + & \text{ethene} \\ C_{12}H_{26}(g) & \rightarrow & C_{10}H_{22}(g) & + & C_2H_4(g) \\ \text{(found in kerosene)} & & \text{shorter alkane} & & \text{alkene} \end{array}$$

It is possible to distinguish between alkanes and alkenes by adding bromine water. With the alkene the orange colour of the bromine water disappears.

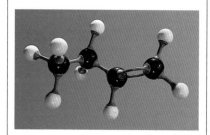
butene

Structure and shape of the first three alkenes

★ THINGS TO DO

1 Estimate the boiling points of the alkanes with formulae:
   a) C_8H_{18} **b)** $C_{12}H_{26}$

2 Write a balanced chemical equation to represent the combustion of propane.

3 Using the information in the table about alkenes, make an estimate of the boiling points of hexene (C_6H_{12}) and octene (C_8H_{16}).

4 Write a balanced chemical equation to represent the process which takes place when decane ($C_{10}H_{22}$) is cracked.

5 Write a word and balanced chemical equation for the reaction between ethene and hydrogen chloride.

6 Write the structural formula for pentene.

Giant molecules

In 1933, Gibson and Fawcett carried out a reaction involving ethene and another organic chemical called benzaldehyde at a high pressure of about 2000 atmospheres. The reaction vessel leaked and some air (oxygen) got in. More ethene had to be added. When they opened the reaction vessel they found a white waxy solid – quite different to the substance they had expected. They had discovered, by accident, a completely new type of substance.

This new substance had very large molecules, each one made up of many thousands of ethene molecules which had bonded together (see illustration). The Greek word for 'many' is 'poly' and so the new substance was given the name polyethene (or **polythene**). The modern plastics industry was born with this accidental discovery.

Polythene has many useful properties:

- it is easily moulded;
- it is an excellent electrical insulator;
- it does not corrode;
- it is tough;
- it is not affected by the weather;
- it is durable.

It was first used to insulate telephone cables and its unique property as an electrical insulator was essential during the development of radar. Its properties continue to be exploited and today it can be found as a substitute for natural materials in plastic bags, sandwich boxes, washing up bowls, wrapping film, milk bottle crates and squeezy bottles (see photograph).

Modern processes for manufacturing polythene

We now manufacture polythene by heating ethene to a relatively high temperature, under a high pressure in the presence of a catalyst.

$$n \begin{pmatrix} \begin{matrix} H & & H \\ \diagdown & & \diagup \\ & C = C & \\ \diagup & & \diagdown \\ H & & H \end{matrix} \end{pmatrix} \longrightarrow \begin{pmatrix} \begin{matrix} H & H \\ | & | \\ -C - C - \\ | & | \\ H & H \end{matrix} \end{pmatrix}_n$$

where n is a very large number.

In polythene the ethene molecules are joined together to form a very long hydrocarbon chain

The ethene molecules join together very much like these beads. Each bead represents a single ethene molecule. The connecting pieces represent the bonds which link the ethene molecules together into one long string

Polythene has many uses

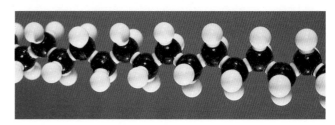

This shows part of the polythene polymer chain

(see illustration). The ethene molecules are able to form chains like this only because they possess carbon-carbon double bonds.

Other alkene molecules can also produce substances like polythene. Propene, for example, produces polypropene which is used to make ropes and packaging.

When small molecules like ethene join together to form long chains of atoms, the process is called **polymerisation**. The small molecules, like ethene, which join together in this way are called **monomers**.

A **polymer** chain often consists of many thousands of monomer units and in any piece of plastic there will be many millions of polymer chains. Since in this polymerisation process the monomer units add together to form the polymer the process is called **addition polymerisation**.

Other addition polymers

Many other addition polymers have been produced. Often the plastics are produced with particular properties in mind, for example, PTFE (polytetrafluoroethene) and PVC (polyvinyl chloride or polychloroethene). Both of these plastics have monomer units similar to ethene.

PVC monomer
(vinyl chloride or chloroethene)

PTFE monomer
(tetrafluoroethene)

If we start from chloroethene, the polymer we make is slightly stronger and harder than polythene and is therefore particularly good for making pipes for plumbing.

monomer → polymer chain

If we start from tetrafluoroethene, the polymer we make has some slightly unusual properties:

• it will stand very high temperatures
• it forms a very slippery surface.

These properties make PTFE an ideal 'non-stick' coating for frying pans.

monomer → polymer chain

The properties of some addition polymers along with their uses are given in the table.

Properties of addition polymers

Plastic	Monomer	Properties	Uses
Polythene	$CH_2\!=\!CH_2$	Tough, durable	Carrier bags, bowls, buckets, packaging
Polypropene	$CH_3CH\!=\!CH_2$	Tough, durable	Ropes, packaging
PVC	$CH_2\!=\!CHCl$	Strong, hard (less flexible than polythene)	Pipes, electrical insulation, guttering
PTFE	$CF_2\!=\!CF_2$	Non-stick surface, withstands high temperature	Non-stick frying pans, soles of irons
Polystyrene	$CH_2\!=\!CHC_6H_5$	Light, poor conductor of heat	Insulation, packaging (especially as foam)
Perspex	$CH_2\!=\!C(CO_2CH_3)CH_3$	Transparent	Used as a glass substitute

---

★ THINGS TO DO

1 Use the information in the table to answer the following questions:
a) Draw out the structure of the monomer propene.

b) Draw a diagram to show a section of the polymer formed from propene.
c) Draw the structure of the polymer called perspex.

Exam questions

denotes higher level questions

1   Diagram 1 shows where scientists think the positions of the continents were 200 million years ago. Diagram 2 shows the present-day positions of the continents.
a) Four land masses have been labelled A, B, C and D on Diagram 1. Use the same letters, A, B, C and D, **to label on Diagram 2** the present day positions of the four land masses.   (2)
b) Scientists think that the continents have moved apart over the last 200 million years. Give **two** pieces of evidence for the movement.   (2)
c) The crust of the Earth includes several tectonic plates. What causes tectonic plates to move?   (3)
d) The tectonic plate labelled D in diagram 1 eventually collided with an oceanic tectonic plate. Describe what happens to tectonic plates when they collide.   (4)

(NEAB, Specimen)

Diagram 1

Diagram 2

2   **a)** The main gases in the Earth's atmosphere are nitrogen, oxygen, water vapour and carbon dioxide. When the atmosphere was first formed it was mainly composed of carbon dioxide, nitrogen, methane and water vapour.
The table below shows how the percentages of some gases in the atmosphere have changed over time.

i) How has the percentage of carbon dioxide in the atmosphere changed?   (1)
ii) Suggest TWO reasons for this change.   (2)
b) Explain how the gradual evolution of different life forms is related to changes in the atmosphere.   (5)

(ULEAC, 1994)

Millions of years ago	Percentage composition			Main events
	Carbon dioxide	Nitrogen	Oxygen	
4500	90	10	0	Earth formed
4000	40	30	0	oceans formed
3500	21	40	trace	
3000	15	55	1	first sea plants
2500	10	60	5	
2000	7	70	10	
1500	5	75	18	
1000	2	77	20	first sea animals
500	1	77	21	first land plants and animals
0	less than 1	78	21	

3   The diagram shows some features of the Earth's crust.
a) List the **four layers** of sedimentary rock named in the diagram, in order of age, starting with the oldest.   (1)
b) i) What types of rocks are likely to be formed in region **X**?   (1)
ii) Name a rock formed from limestone in region **X**.   (1)

Ground surface

Igneous rock

Region X

Sandstone A
Shale
Limestone
Sandstone B

c) Describe how sedimentary rock, such as sandstone, is formed from sediment. (2)

d) The sketch map **A** shows a region in Scotland. Two granite areas are shaded in.
It is thought that the granite once formed part of a single area (sketch map **B**).
Separation along a fault line has continued over a period of time.

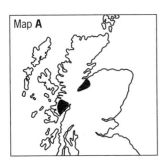 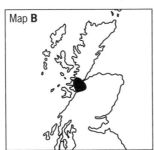

i) Draw on map **A** the fault line responsible for the separation. (1)
ii) Add arrows to map **A** to show the direction of movement on **each side** of the fault. (1)

e) The granite found in both areas contained crystals which were much smaller than those found in another location. Explain why the crystal size differs. (2)

(SEG, 1994)

4 A team of geologists studied an island. The team was most interested in the rocks found in the four areas marked **A**, **B**, **C** and **D** on the diagram below.

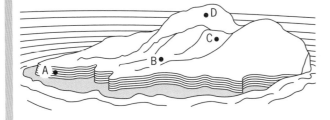

The table below shows part of their report. Use the information in it to help you to answer the following questions.

Area of rock sample	Is it made up of layers?	Is it hard?	Does it have crystals in it?	Does it fizz if acid is added?
A	yes	no	no	yes
B	yes	yes	no	no
C	no	yes	yes, large ones	no
D	no	yes	yes, small ones	yes

a) Rocks are often identified by using a key. Write, in the diamonds in the key below, the questions which will lead to the correct rock areas being identified. (3)

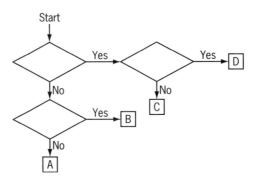

b) Which area of rock sample is most likely to have been formed from hot molten rock which cooled slowly? State the reason for your choice. (2)

c) Rock samples from **A** and **D** are two forms of calcium carbonate.
i) Which gas is given off when the acid is added? (1)
ii) Complete the gaps in the following sentences. (3)
To test for this gas a solution can be used. This solution is called ....
This solution would turn from being ....
to .... if this gas is present.

d) The rock sample from **D** is a metamorphic rock. Describe **two** ways metamorphic rocks can be formed. (3)

(SEG, Specimen)

5 The diagram shows how crude oil can be broken down into fractions. Each of these fractions is a useful material.

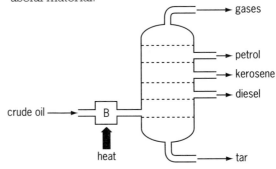

a) i) What is the **full name** of this process? (2)
ii) What change to the crude oil takes place in the part labelled **B**? (1)

b) Some of the fractions are fuels.
i) How is the energy released from a fuel? (1)
ii) What else is produced when the energy is released? (1)

c) Name **ONE** polymer which is made from crude oil. (1)

(MEG, 1995)

6 a) The table below gives some information about the group of hydrocarbons called the alkanes.

Name	Formula	Melting point (°C)	Boiling point (°C)	Density (g/cm³)	Energy released by combustion (kJ/mol)
Methane	CH_4	−182	−162	0.53	890
Ethane	C_2H_6	−183	−89	0.55	1560
Propane	C_3H_8	−188	−42	0.57	2220
Butane	C_4H_{10}	−138	−1	–	2880
Pentane	C_5H_{12}	−130	+36	0.63	–
Hexane	–	−95	+69	0.66	4200

i) What is the pattern linking density and the amount of heat energy released by combustion? (1)
ii) Suggest a value for the density of butane. (1)
iii) What pattern is shown by the formulae of the alkanes? (2)
iv) Use the pattern to predict the formula of hexane. (1)

b) The graph below shows how the energy released by combustion (*y* axis) changes with the number of carbon atoms (*x* axis) for the alkanes.

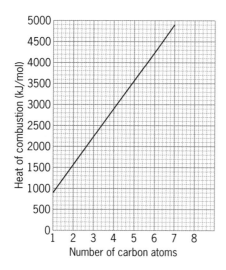

i) Use the graph to estimate the energy released by combustion of pentane. (1)
ii) What part of the air is needed for the pentane to burn? (1)
iii) What TWO substances are formed when pentane is burnt in a good supply of air? (2)
iv) Name another substance that is formed when pentane is burnt in a limited supply of air. (1)
v) The flame formed when hydrocarbons are burned becomes smokier as the number of carbon atoms increases. Suggest a reason why. (1)

(ULEAC, 1995)

7 Crude oil is a mixture of many compounds. Most of these compounds are hydrocarbons. The structure of one of these compounds is shown in the diagram.

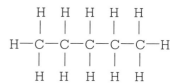

a) What is a hydrocarbon? (1)
b) What is the chemical formula of the structure shown in the diagram? (1)
c) In which crude oil fraction would this compound be found? Use the **Data Book** to help you. (1)
d) Name the method of separation used to obtain the various fractions of crude oil. (2)
e) Cracking is one of the most important processes in the oil industry. Cracking involves the thermal decomposition of large molecules. The diagram below shows an apparatus that can be used to demonstrate cracking in the laboratory. The porous pot acts as a catalyst in the reaction.

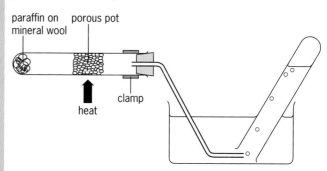

i) What happens to the large molecules during thermal decomposition? (2)
ii) What effect does the catalyst have on the reaction? (1)
f) Ethene is an unsaturated hydrocarbon. What is meant by the term 'unsaturated'? (1)
g) Complete the following equation to show how three ethene molecules join together to form part of a poly(ethene) molecule. (2)

$$H \atop H \rangle C = C \langle {H \atop H} \quad + \quad {H \atop H} \rangle C = C \langle {H \atop H} \quad + \quad {H \atop H} \rangle C = C \langle {H \atop H}$$

i) Suggest one property of poly(ethene) which makes it suitable as a material for food containers. (1)

ii) Thermosetting plastics cannot be re-moulded after they have been heated and allowed to cool. Explain why. (2)

(NEAB, Specimen)

STRUCTURE AND BONDING

Solids, liquids and gases

Some of the things around you are **solid**. Others are **liquid**. Yet others are **gas**. Of the thousands of substances you know, they will either be solid, liquid or gas. Solid, liquid and gas are the three states in which all substances exist. They also have quite different properties which affect the way they behave and the way they are used.

Most solids are hard, strong and keep their shape unless large forces act on them. Some are stronger than others. Railway lines must keep their shape otherwise trains would be derailed and bridges must be able to withstand large forces.

Liquids are runny – they can flow from place to place. Petrol is a liquid which flows through pipes from the petrol tank to the engine in a vehicle and water flows through many miles of pipes to reach your home. If a liquid is put into a container it fills the space – the liquid takes up the shape of the container, although the volume does not change.

Gases can also flow from place to place. They have no shape or volume but fill up the space in whatever they are contained. Unless kept in a closed container, they spread out quickly.

Right Bridges often have steel rods inserted into the concrete to add extra strength

Below left When poured a liquid's shape changes, but the volume does not

Below right The gas fills the space inside the castle making it rigid. It is easily compressed (squashed) and so protects the children bouncing on it

Particle ideas

All **matter** is made up from millions of **particles**, far too small to be seen. The way the particles are arranged helps us to understand some of the properties of solids, liquids and gases that have just been described.

In a solid each particle attracts others around it. These forces of attraction hold the particles together quite strongly. The particles are locked in position, arranged in a regular way. Although they do not move out of position, they do vibrate to and fro. The higher their temperature, the faster they vibrate.

The particles in a liquid are also close together. They also attract one another but the forces are weaker than those between the particles in a solid. The forces are not strong enough to hold the particles in position, so they slide around one another quite freely. They move around quite quickly and randomly, often colliding with other particles.

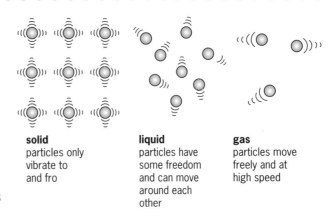

solid
particles only vibrate to and fro

liquid
particles have some freedom and can move around each other

gas
particles move freely and at high speed

The arrangement of particles in the three **states of matter**

In a gas the particles are much further apart than in a solid or a liquid. They are free to move anywhere inside the space they occupy. They move randomly, in all directions, at very high speeds (much higher speeds than those in a liquid). This is why gases spread out (diffuse) and mix completely with each other. They collide with one another but, because they are further apart, collide less often than the particles in a liquid.

★ THINGS TO DO

1 Copy and complete the table below.

Material	Solid, liquid or gas	Property which makes it useful for this job
Air in bicycle tyres		
Petrol for cars		
Wood for chairs		
Steel tubing for bicycles		

2 Oils are liquids which are used to reduce **friction** in machinery. Plan a test to find out how the runniness (viscosity) of oil is affected by its temperature. Think carefully about the things you will need to do to make your tests safe.

When your teacher has approved your plans, carry out your tests and prepare a report.

3 Which of the cartoons showing students in the playground, during lessons and between lessons resemble how particles behave in a solid, a liquid and a gas most closely?

Changing state

Some substances exist naturally in all three states – solid, liquid and gas. Under normal conditions, whether something appears as a solid, liquid or gas depends on its temperature. Water, for example, freezes when its temperature falls to, or below, 0°C. This temperature is the **melting point** of water. Above 100°C – the **boiling point** of water – it exists as a gas – steam. Between these two temperatures it is only found as a liquid.

On Earth water exists in all three states of matter

At temperatures at or above its boiling point, a substance will exist as a gas.

boiling point 100° C

At temperatures between the melting point and boiling point a substance appears in its liquid state

melting point 0° C

At temperatures at or below the melting point a substance appears in its solid state.

Changes of state

Although we have shown water as an example, all other substances behave in the same way.

Using changes of state

Solder is a solid at normal temperatures. It is melted by the heat of the soldering iron to make good contact between the metals being joined

Special effects on stage are often created using solid carbon dioxide. Solid carbon dioxide turns straight into a gas when it is warmed – a process called **sublimation**

What happens to the particles?

To change the state of a substance it must be heated or cooled – it is the temperature that decides whether it will be solid, liquid or gas.

The substance changes state due to the behaviour of the particles in it.

When a solid is heated the particles gain energy and vibrate faster. Eventually the particles break free from one another and the regular pattern of the structure breaks down. The solid has melted. The temperature at which this happens is the melting point.

If the liquid cools down its particles lose energy. As this happens the particles vibrate less. At a certain temperature the forces of attraction between the particles pull them back to the regular arrangement of the solid. The temperature at which the substance solidifies (changes into a solid) is the same as the melting point.

If the liquid is heated the particles move even faster as their average energy increases and they move further apart. At all temperatures some have enough energy to

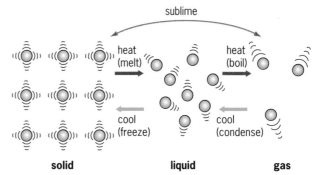

Summary of the changes of state

overcome the forces of attraction of other particles and escape to form a gas – they evaporate.

At the boiling point the liquid particles change into gas so quickly that bubbles of gas rise from inside the liquid. The temperature of a liquid cannot be higher than its boiling point except under special conditions.

When the gas is cooled the average energy of the particles decreases and they slow down. The forces of attraction between them draw the particles closer together causing the gas to condense into a liquid.

★ THINGS TO DO

1 Make a list describing some changes of state which you have seen. For each one describe what happened to bring about the change.

2 James bought a huge balloon at the fair on a hot summer day. Before he went to bed he noticed that the balloon was much smaller. Suggest two possible reasons for the change in shape of the balloon.

3 Petrol becomes completely solid at −50°C. Diesel fuel becomes completely frozen at −20°C. The lowest temperature reached in winter in the United Kingdom was −30°C in Scotland in 1995. The drivers of diesel lorries had to heat the fuel in the tanks to start their engines.
   a) Why did the drivers have to warm the fuel?
   b) Would petrol vehicles have the same problem? Explain your answer.

4 The table below shows the boiling point of some substances.

Substance	Boiling point (°C)
Magnesium oxide	3627
Water	100
Oxygen	−183
Copper	2567
Hydrogen	−253
Ethanol (alcohol)	79

   a) Which substances would be liquids at 200°C?
   b) Which substances would not be gases at 200°C?
   c) Which substance has the strongest forces of attraction between its particles?
   d) Which substance has the weakest forces of attraction between its particles?
   Explain your answers to parts c) and d).

Larger and smaller

Trains were brought to a halt in parts of England during the hot summer of 1995 as railway lines buckled. The buckling was caused by the particles in the steel rails **expanding** (getting further apart) as their temperature rose.

The particles in a railway line show the same regular arrangement as most solid substances. The more they are heated the faster they vibrate and the further they move apart. As the steel expands, each length of track gets longer until eventually it pushes against the neighbouring sections. With no more space to fill, further expansion creates huge forces which cause the track to buckle.

Cooling has the opposite effect on most materials, causing the particles to draw closer together. This shortens the material – it **contracts**.

Heat has caused these rail tracks to buckle

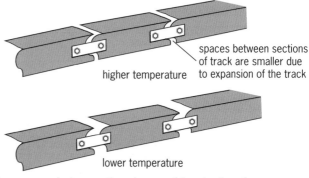

higher temperature   spaces between sections of track are smaller due to expansion of the track

lower temperature

The spaces between the pieces of track allow for expansion and contraction

All substances, whether they are solid, liquid or gas are affected in a similar way by changes in their temperature. They occupy more space (expand) when they are heated and shrink (contract) when cooled.

But try squashing them!

Generally, unless very large forces act on them, solids keep their shape. We have already seen that liquids and gases can take up any shape – they fill the space inside the container in which they are stored so their shape is that of the container. When poured from one container to another their volume does not change.

Electricity cables are deliberately left slack so that they can contract (shrink) during cold spells without snapping

Liquids and gases can be squashed (**compressed**), changing their volume. The illustrations show what happens when different forces act on air trapped inside a syringe. Clearly the gas is compressed – particles have been pushed closer together by the extra pressure of the weights on the plunger. (See *GCSE Science Double Award Physics*, Topic 3.6 on pressure.)

Although liquids can be compressed in the same way, the change in volume is very small compared with that of the gas, even when large forces are applied.

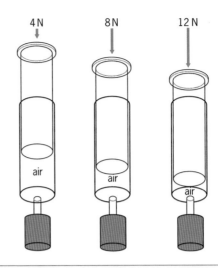

★ THINGS TO DO

1 Use the particle idea of matter to explain why gases can be compressed much more easily than liquids.

2 A gas bottle contains liquefied gas at high pressure. Using the particle model, explain why gases turn to liquid if they are compressed sufficiently.

3 The diagram shows what happens to the volume of air in a bicycle pump when the cylinder is pushed in. Draw what you think the air particles would look like in both pumps. One particle has been drawn in for you.

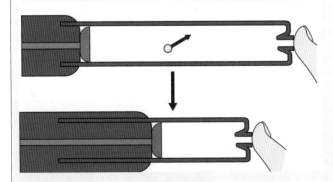

4 The illustration shows a device which opens greenhouse windows as the temperature rises and closes them as the temperature falls.

Write a section for a 'How it works' magazine explaining how the device works.

5 A simple thermometer can be made using cooking oil, a test tube and a length of glass tube.

Plan how you could make a scale for a thermometer such as this. The only clue you can have is that water boils at 100°C and ice melts at 0°C. Oil also expands by the same amount for each °C rise in temperature.

If your teacher is happy with your plan you may be allowed to test it. If so, check the accuracy of your thermometer and think about how it could be made more accurate if necessary.

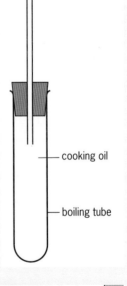

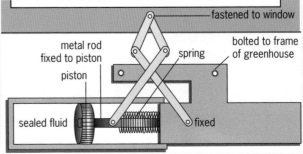

The kinetic theory

Most of the things we have seen so far can be explained by particle ideas. These particle ideas are the basis of the **kinetic theory** – a theory about how particles behave under different conditions. Much of what we have seen supports the kinetic theory which says that:

- all matter is made of very tiny particles which, as we shall see later, are called atoms, **ions** and **molecules**;
- the higher the temperature of a substance the faster its particles move;
- at the same temperature, heavier particles move more slowly than lighter particles.

The kinetic theory can also be used to explain everyday events.

Diffusion

Many people now use pot-pourri – dried, scented leaves – to make their homes smell nice. Some use solid blocks of air freshener; yet others sprinkle powder on carpets before hoovering, or use aerosol air fresheners. Pleasant smells spread through the home from each one.

In each case the smell spreads by a process called **diffusion**. The spreading out takes place in a haphazard, or **random** way, as the gas molecules spread out to fill the space available to them. As the particles spread out they will collide with the walls, and with each other, each time bouncing off in a different direction.

There are many ways in which diffusion can be demonstrated in the laboratory. These photographs show what happens to an orange-red gas called bromine which was placed in the bottom gas jar.

 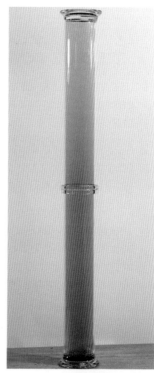

After 24 hours the bromine fumes have diffused throughout both gas jars

Pleasant smells spread through your house from each of these

After 24 hours the bromine fumes have spread evenly throughout both jars. It is the natural movement of the particles of bromine gas which causes the gas to spread out – no draughts can get into the jars to spread the gas.

The particles of some gases are heavier than others and move at different speeds. It seems sensible to assume that heavier, slower particles will diffuse more slowly than lighter, faster particles. This can be shown to be the case using a long glass tube in which hydrogen chloride and ammonia gases can spread.

The hydrogen chloride gas reacts with ammonia gas where they meet, forming a white cloud of ammonium chloride. You can see from the picture that the gases do not meet at the centre, but towards the hydrogen chloride end of the tube. This is because the

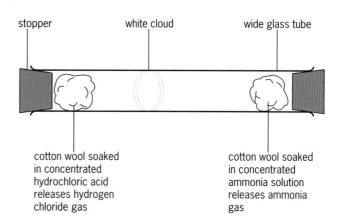

Diffusion of hydrogen chloride and ammonia

hydrogen chloride particles are larger than the ammonia particles, and move more slowly through the tube. They do not, therefore, travel as far as the ammonia particles in the same time.

★ THINGS TO DO

1 At room temperature gas particles move at speeds similar to the speed of sound – about $300 \, m \, s^{-1}$. If they move at this speed, why don't you smell food much faster than you do?

2 The average speed of any moving object, including small particles of gas, can be calculated using the equation:

$$\text{speed} = \frac{\text{distance}}{\text{time}}$$

Plan how you could calculate the average speed at which onion particles (or any other smelly substance) travel through the air in the laboratory.

3 The diagram shows the times and distances measured when ammonia and hydrogen chloride gases diffused in a long tube.

a) Calculate the speed of:
i) the particles of ammonia gas;
ii) the particles of hydrogen chloride gas.
b) What can you tell from the speed of each particle?

4 Each year many people die from carbon monoxide poisoning due to poorly serviced oil heaters. Under these conditions the fuel does not burn properly producing poisonous carbon monoxide gas. Many of the victims may not be in the same room as the heater. Explain how the carbon monoxide gas can move from room to room in a house.

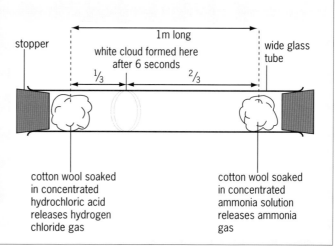

Warming up and cooling down

Warm air from the hand drier speeds up the evaporation of water from the hands

Whenever a change of state occurs, energy is transferred. The energy changes can be summarised as:

As the air warms during the day, icicles melt

Change of state	What happens	What the energy change causes
solid→liquid	Energy transferred to solid, heating it	Breaks forces of attraction between solid particles
liquid→gas	Energy transferred to liquid, heating it	Breaks forces of attraction between liquid particles
gas→liquid	Energy transferred from gas, cooling it	Particles slow down; forces of attraction draw particles together
liquid→solid	Energy transferred from liquid, cooling it	Particles slow down further; forces of attraction draw particles into solid state

Evaporation and boiling

While you are having a bath the room often fills with water vapour – what you might call steam. The vapour must come from the water in the bath, yet the water is not boiling. Some liquid water particles must turn to steam at temperatures lower than the boiling point. This is evaporation.

Evaporation (change of state from liquid to gas) can occur at any temperature. Evaporation dries the ground after rain. Evaporation dries you after swimming. Some of the liquid in perfumes, aftershaves and correction fluid dries up due to evaporation. So how do liquids evaporate?

In all liquids the particles are moving. At higher temperatures they move faster. As they move in the liquid they collide with other particles, and with the walls of the container. During these collisions, some particles gain energy (at the expense of those they collide with). They gain sufficient energy to 'tear' themselves away from the forces of attraction of other particles. Close to the surface they leave the liquid, becoming gas particles. This is evaporation and happens at all temperatures.

As the liquid is heated, the particles move faster. More of them are able to gain sufficient energy to break free from the liquid becoming gas particles. Evaporation is therefore faster at higher temperatures.

As we continue heating the liquid it eventually reaches its boiling point. At this temperature particles throughout the liquid change to gas, and bubbles can be seen rising through the liquid.

This is boiling. Boiling occurs only at the boiling point of the particular liquid.

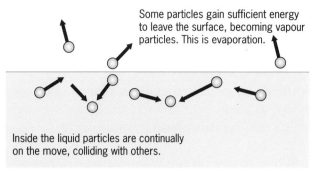

Some particles gain sufficient energy to leave the surface, becoming vapour particles. This is evaporation.

Inside the liquid particles are continually on the move, colliding with others.

Evaporation takes place in this way

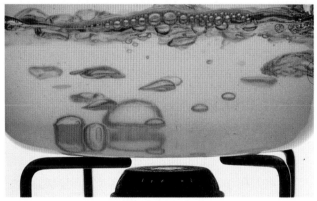

At the boiling point bubbles of gas can be seen forming throughout the liquid

★ THINGS TO DO

1 This diagram shows the way in which the liquid in a refrigerator (the refrigerant) flows around a system of pipes which pass inside and outside the fridge.

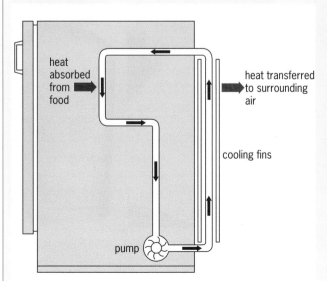

heat absorbed from food

heat transferred to surrounding air

cooling fins

pump

a) Copy the picture into your notebook. Add your own labels showing where:
i) the refrigerant will change from the liquid to the gaseous state;
ii) the refrigerant changes from the gaseous state to the liquid state.
b) Add colours to the pipes on the diagram showing where the refrigerant will be:
i) liquid;
ii) gas.
c) Why do the fins at the back of the refrigerator get hot?

2 Make a list of factors which you think affect how quickly a liquid evaporates. For each one give a reason for your prediction. Plan some tests which you could use to test your ideas, thinking particularly about safety.

Carry out your tests when your teacher has approved your plan, and write a full report describing your findings.

The solution

Tablets – no thanks!

Some people cannot swallow tablets. To make medicines easier to swallow, many manufacturers now produce tablets which can be **dissolved** in water, such as the **soluble** aspirin shown here.

When stirred in water the tablet dissolves, leaving no trace of solid. The solid aspirin has dissolved in the water to form a **solution**. The solid chemicals in the tablets are the **solute** and water is the **solvent**. The solute dissolves in the solvent to form a solution. The process of dissolving involves the separation of the added solute particles as the liquid particles move about and collide with them.

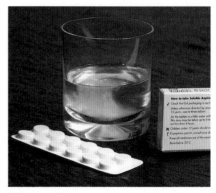
This medicine can be dissolved in water

Faster energy

Solids cannot be absorbed into the blood stream. They must first be dissolved. Isotonic drinks are popular with sporting people because they are a quick source of energy. Drinks such as these are solutions. The solutes are solids such as glucose (a type of sugar) dissolved in water. Carbon dioxide gas is also dissolved to produce the fizz. The sugar in the solution is absorbed from the small intestine into the blood stream more quickly than if taken as a solid, releasing energy quickly.

Solutions are always clear (there are no solid particles to be seen) although they may be coloured. Some isotonic drinks, for example, are yellow or orange. If the solute is yellow, then the solution will be yellow. When dissolved the particles of the solute spread evenly throughout the solvent and a mixture is created.

salt just after being placed in water

solution a short time after mixing

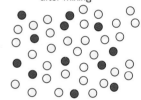

Salt particles hold one another together, initially. Water particles begin to pull salt particles apart.

Soon the salt particles are scattered through the solvent – we have a solution. The salt has dissolved in the water to form a salt solution.

A solute dissolving in a solvent to form a solution

Other solvents

Water is not the only solvent which can be used to dissolve substances. There are many solutions which we use every day which use other solvents.

Some solvents are dangerous. When the solvent evaporates it produces fumes which can be poisonous or flammable or both.

Athletes need energy fast

Correction fluid is a solution of a plastic dissolved in non-toxic, non-volatile aliphatic hydrocarbons

Aftershave is a solution of plant extracts dissolved in ethanol

Soluble or insoluble?

In the examples described on the previous page a solid was dissolved in a solvent to produce a solution. The solids are described as soluble because they dissolve in the solvent (hence the term 'soluble aspirin'). A substance can, however, be soluble in one solvent but not in another.

At some time you may have had chewing gum stuck to your clothes. Because chewing gum will not dissolve in water, it cannot be easily removed by washing. It does, however, dissolve in a solvent called xylene. As it dissolves, the clothes are left unmarked. Unfortunately xylene is one of those solvents which produce harmful vapours as they evaporate.

The chewing gum is quite soluble in xylene

Suspensions

Milk of magnesia is made by mixing insoluble magnesium hydroxide with water. If it is left to stand the magnesium hydroxide falls to the bottom of the bottle. When you shake the bottle containing this mixture you get a suspension. The solid particles remain suspended in the water until they settle to the bottom again under the force of gravity. Other examples of this type of mixture include calamine lotion which is used for itchy spots like those you get with chickenpox.

This indigestion remedy is a suspension

★ THINGS TO DO

1 Many medicines are prepared by dissolving a solid form of the medicine in a solvent such as glucose solution. The pharmacist needs to know how to dissolve the solids quickly.

Make a list of factors which could affect how quickly a solute dissolves in a solvent. Give a reason for each one.

Test your ideas after they have been approved by your teacher and prepare a report in a form which would be helpful to the pharmacist.

2 Why do you think medicine bottles usually advise you to 'shake well before use'?

3 A science book for younger children said that when sugar was put into water it 'disappeared'. The sugar does not disappear – it is still there but cannot be seen. Draw a series of pictures aimed at younger children which would provide a better description of dissolving.

4 Describe how you would show someone that Lucozade was a solution containing a solid dissolved in a liquid.

Particles

In the previous topics we have used the word 'particle' quite often – almost as if the particles of all substances were the same. In fact there are three main types of particle – atoms, molecules and ions – although you will see that the atom is the basic particle from which other things are made.

What is an atom?

Everything is made of atoms – including you! They are far too small to be seen. For many years it was thought that the atom was the smallest particle of matter which could exist. Now we know that even atoms are made up of many smaller particles – sub-atomic particles.

Different substances are made up of different atoms. The atoms of gold, for example, are only slightly different to those of

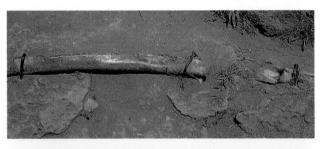

You cannot change lead (Roman pipe, *top*) into gold jewellery

lead, although the two substances themselves are quite different, both in appearance, their properties, and their value! For many years people tried to find a way of turning lead into gold – unsuccessfully.

Elements are made of only one type of atom. Iron contains nothing but iron atoms. Silver contains nothing but silver atoms. Again there are only slight differences between the atoms, but the differences are enough to give them quite different properties.

There are only 92 different types of naturally occurring atom. Others, some of which exist for only fractions of a second, have been made in laboratories. These 92 atoms combine in millions of different ways to form millions of substances.

Molecules

Although some substances can exist as single atoms, most contain groups of atoms held strongly together by strong attractive forces, or **bonds**. The groups may be quite small or can be very large, such as those found in some plastics. These groups of atoms are called molecules.

The oxygen we breathe consists of pairs of oxygen atoms held together by bonds. Water, one of the commonest substances on Earth, is made up of molecules containing two hydrogen atoms bonded to a single oxygen atom. Molecules are often shown in picture form.

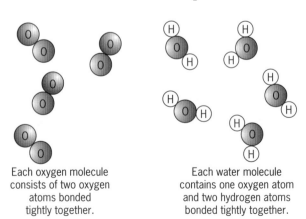

Each oxygen molecule consists of two oxygen atoms bonded tightly together.

Each water molecule contains one oxygen atom and two hydrogen atoms bonded tightly together.

Oxygen molecules   Water molecules

The numbers of each atom present can be represented as a formula. Water, for example, has two molecules of hydrogen joined to one atom of oxygen. We can write this as a formula – H_2O (see page 117 for more information on formulae).

Oxygen is an element – a substance containing only one type of atom. Water, on the other hand, is a compound – it contains two different types of atom. Compounds are substances which contain two or more different types of atom bonded together. With the exception of the elements, every other substance on Earth is a compound. The glucose seen in the drink in the previous topic is a more complicated compound – it has 6 atoms of carbon, 6 atoms of oxygen and 12 atoms of hydrogen, all of which are held strongly together by chemical bonds.

Ions

Atoms are uncharged particles. An ion is a particle which carries a small electric **charge** – we say it is a charged particle. The charge can be positive or negative (see *GCSE Science Double Award Physics*, Topic 2.1). Most ions are formed when atoms undergo changes in their structure, as you will see in the next unit.

Oppositely charged objects attract one another (see *GCSE Science Double Award Physics*, Topic 2.1) so you might expect a sodium ion to be strongly attracted to a chloride ion. It is that force of attraction which holds these two particles together in salt – a compound called sodium chloride.

The oppositely charged ions hold one another together in a regular shape. Ions with the same charge would repel one another – they cannot hold one another together. It follows that substances which contain ions must contain both positive and negative ions, and so must always be compounds.

The formula of glucose is $C_6H_{12}O_6$

a sodium atom   a positively charged sodium ion

a chlorine atom   a negatively charged chloride ion

Metals always form positive ions. Non-metals usually form negative ions

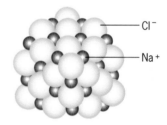

Sodium and chloride ions held together in a lattice

★ THINGS TO DO

1 How many atoms of the different elements are there in the formula of each compound given below:
   a) methane (Natural gas), CH_4;
   b) sulphuric acid, H_2SO_4;
   c) ethanoic acid (found in vinegar), CH_3COOH;
   d) chloroform, $CHCl_3$;
   e) propanone (a glue solvent), CH_3COCH_3.

2 What is meant by the terms:
   a) atom;
   b) element;
   c) compound;
   d) molecule;
   e) ion.
   Use an example to help you with your explanation.

3 Potassium chloride, used as a substitute for salt, contains ions.
   a) How many elements are present in this substance?
   b) Which of the elements present form a negative (−) ion and which a positive (+) ion?
   c) Draw a neat diagram to show the lattice formed by the ions present in this substance.

What's in an atom?

About 100 years ago scientists still believed that atoms were solid particles like very tiny marbles.

Since then a lot of evidence has been produced by scientists such as Bohr, Moseley, Thomson, Einstein, Rutherford and Chadwick which shows that all atoms are made up of three tiny particles called **protons, neutrons** and **electrons**.

The protons and neutrons are found in the centre of the atom, which is called the **nucleus**. The nucleus occupies only a very small volume of the atom but it is very dense. The neutrons have no charge and the protons are **positively charged**.

The rest of the atom surrounding the nucleus is where the electrons are most likely to be found. The electrons are **negatively charged** and move around very quickly in **electron shells** or **energy levels**. The electrons are held within the atom by an **electrostatic force of attraction** between themselves and the positive charge of the protons in the nucleus.

Protons and neutrons have approximately the same **mass** – but far greater than the mass of an electron. About 1837 electrons would have the mass of one proton or neutron. A summary of the characteristics of each type of particle is given in the table.

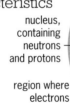

nucleus, containing neutrons and protons

region where electrons are found

The smaller particles in an atom

Characteristics of the sub-atomic particles in an atom

Particle	Symbol	Relative mass (amu)	Relative charge
proton	p	1	+1
neutron	n	1	0
electron	e	1/1837	−1

amu = atomic mass unit

You will notice that the masses of all three particles are measured in **atomic mass units (amu)**. This is because they are so light that their masses cannot be measured usefully in grams.

Atoms are neutral (uncharged) particles

The atoms themselves are electrically **neutral**. This is because they contain equal numbers of protons and electrons. The positive charge on the protons effectively cancels the negative charge on the electrons.

In an atom of carbon, for example, there are 6 protons, 6 neutrons and 6 electrons. The (positive) electrical charge of the protons in the nucleus is balanced by the opposite (negative) charge of the 6 electrons.

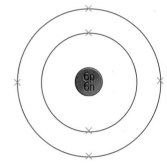

An atom of carbon has 6 protons, 6 neutrons and 6 electrons

Proton number (atomic number) and mass number (nucleon number)

The number of protons in the nucleus of an atom is called the **proton number** (or the **atomic number**) and is given the symbol Z. The carbon atom has an atomic number of 6, since it has 6 protons in the nucleus. Each element has a different proton number.

Neutrons and protons have similar masses. Electrons possess very little mass in comparison. The mass of an atom must depend largely on the number of protons and the neutrons in its nucleus.

The total number of protons and neutrons found in the nucleus of an atom is called the **mass number** (or **nucleon number**) and is given the symbol A.

$$\text{mass number}(A) = \text{proton number}(Z) + \text{number of neutrons}$$

Hence in the case of carbon shown in the illustration on the previous page, it has a mass (nucleon) number of 12, since it has 6 protons and 6 neutrons in its nucleus.

The proton number and the mass number of an element are usually written in the following shorthand way:

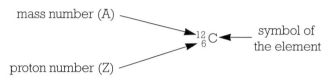

mass number (A) —

$^{12}_{6}C$

symbol of the element

proton number (Z) —

The number of the neutrons can be calculated by rearranging the relationship between the proton number, mass number and number of neutrons to give:

$$\text{number of neutrons} = \text{mass number(A)} - \text{proton number(Z)}$$

The table shows the number of protons, neutrons and electrons for some common elements

Number of protons, neutrons and electrons for some common elements

Isotopes

Atoms of the same element which have different numbers of neutrons are called **isotopes**. There are two isotopes of chlorine, shown in the diagram.

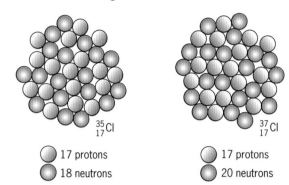

$^{35}_{17}Cl$ $^{37}_{17}Cl$

17 protons   17 protons
18 neutrons   20 neutrons

Generally, different isotopes of the same element behave in the same way during chemical reactions. The only effect of the extra neutrons is to alter the mass of the atoms and the properties which depend on it, such as density. Some of the atoms of certain isotopes are unstable and radioactive because of the presence of the extra neutrons. The best known elements which have radioactive isotopes are carbon and uranium.

Element	Symbol	Proton number	Number of protons	Number of electrons	Number of neutrons	Mass number
Hydrogen	H	1	1	1	0	1
Helium	He	2	2	2	2	4
Carbon	C	6	6	6	6	12
Nitrogen	N	7	7	7	7	14
Oxygen	O	8	8	8	8	16
Fluorine	F	9	9	9	10	19
Neon	Ne	10	10	10	10	20
Magnesium	Mg	12	12	12	12	24
Iron	Fe	26	26	26	30	56

★ THINGS TO DO

1 What is meant by the terms proton number and nucleon number?

2 What are the proton and nucleon numbers of:
a) oxygen;   **b)** magnesium;   **c)** iron?

3 a) Write your answers to question 2 in the chemical shorthand used to show the symbol as well as the proton and nucleon number.
b) Calculate the number of neutrons in each of the atoms in question 2.

Electron arrangement

The electrons which orbit the nucleus of any atom are not randomly arranged, but orbit the nucleus in a particular way. Some, for example, orbit further from the nucleus than others. The position of the electrons, and in particular the number which are in each orbit, significantly affect the way the material behaves.

The nucleus of an atom contains the heavier sub-atomic particles, the protons and the neutrons. The electrons, the lightest of the sub-atomic particles, move around the nucleus at great distances from the nucleus relative to their size. They move very fast in electron energy levels (or shells) very much as the planets orbit the Sun.

It is not possible to give the exact position of an electron in an energy level. However, we can state that electrons can only occupy certain, definite energy levels and that they cannot exist between them. Additionally, each of the electron energy levels can only hold a certain number of electrons.

Although this picture shows only three energy levels, there are others beyond them which contain increasing numbers of electrons.

The electrons fill the energy levels starting with the energy level nearest the nucleus. Electrons in this first energy level have less energy than those in the second. When this level is full (with 2 electrons) the next electron goes into the second energy level. When this energy level is full with 8 electrons, then the electrons begin to fill the third energy level and so on. The third energy level can be occupied by a maximum of 18 electrons. However, when 8 electrons have occupied this level a certain stability is given to the atom and the next 2 electrons go into the fourth energy level.

For example, the common form of oxygen, $^{16}_{8}O$ atom has a proton number of 8 and so has 8 protons and 8 electrons. Two of the 8 electrons enter the first energy level, leaving 6 to occupy the second energy level as shown in the diagram.

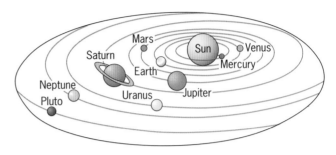

The way electrons orbit a nucleus is similar to the way the planets orbit the Sun

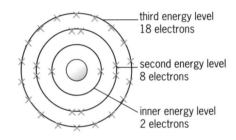

The inner energy level can hold up to 2 electrons. The next can hold 8 electrons and the third 18 electrons

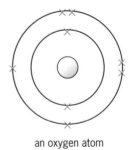

an oxygen atom

The arrangement of the electrons in the $^{16}_{8}O$ atom

The way the electrons are arranged is called the **electron structure** or **electron configuration**. The electron structure of the oxygen atom can be written as 2,6.

There are 112 elements, and the table shows the way in which the electrons are arranged in the first 20 of these elements.

The diagram shows how the electrons are arranged in the energy levels of some of the atoms shown in the table.

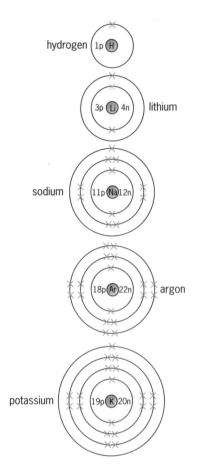

The number of protons, neutrons and electrons in some elements

Element	Symbol	Proton number	Number of electrons	Electron structure
Hydrogen	H	1	1	1
Helium	He	2	2	2
Lithium	Li	3	3	2,1
Beryllium	Be	4	4	2,2
Boron	B	5	5	2,3
Carbon	C	6	6	2,4
Nitrogen	N	7	7	2,5
Oxygen	O	8	8	2,6
Fluorine	F	9	9	2,7
Neon	Ne	10	10	2,8
Sodium	Na	11	11	2,8,1
Magnesium	Mg	12	12	2,8,2
Aluminium	Al	13	13	2,8,3
Silicon	Si	14	14	2,8,4
Phosphorus	P	15	15	2,8,5
Sulphur	S	16	16	2,8,6
Chlorine	Cl	17	17	2,8,7
Argon	Ar	18	18	2,8,8
Potassium	K	19	19	2,8,8,1
Calcium	Ca	20	20	2,8,8,2

Chemical reactions between different substances involve an exchange, or a sharing, of the electrons in the highest energy level of the atoms. Knowing about the electron structure will often help you predict the behaviour of different substances during reactions between them and other substances.

The electron structures of helium (He), neon (Ne) and argon (Ar) are particularly stable since they all have completely filled energy levels. This means that these gases are very unreactive and because of this they are known as the inert gases. (Another name for them is the noble gases.)

★ THINGS TO DO

1 Draw energy level diagrams for the following elements: **a)** carbon; **b)** sulphur; **c)** calcium.

2 From the first 20 elements name and give the symbols of those elements:
a) with 3 electrons in their outer energy levels;
b) which have full outer energy levels;
c) which have three electrons short of a full outer energy level.

3 In what ways are the electron structures of carbon and silicon similar?

4 Explain the following statements.
a) The electrons which are present in an atom are arranged in energy levels.
b) The electron structure of chlorine is 2.8.7.

Well held

The properties of a material depend very much on how the atoms or molecules are arranged inside it – its structure. Slight changes in the structure can alter the properties significantly. How the atoms or molecules are arranged depends on their electron structure – the way the electrons are arranged in each atom or ion.

Different materials from the same atoms

Diamond is the hardest naturally occurring substance known – so hard it is used to drill through rocks in the Earth's crust. It is also, of course, a beautiful (and expensive) gemstone used in jewellery.

Graphite is one of the softest substances. Surfaces covered with graphite slide over one another very easily because the graphite reduces friction – it acts as a **lubricant** (see *GCSE Science Double Award Physics*, Topic 3.9 on friction). Pencil 'lead' is made from graphite.

Surprisingly, both materials contain only carbon atoms yet have quite different properties. That is because the carbon atoms are joined together in different ways in the diamond and the graphite – the materials have quite different structures.

The structure, or the way in which the atoms or molecules are arranged and held together in a material, determines its properties.

Diamonds are spectacular jewels

Pencils contain graphite

a
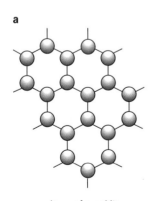

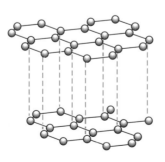

b
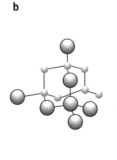

one layer of graphite

showing how the layers fit together in graphite

a small part of the structure of diamond

a view of a much larger part of the structure of diamond

a) The structure of graphite and **b)** the structure of diamond

Bonding

Salt (sodium chloride) is a very important mineral in our diet. Most foods contain it and our diet must contain the right amount of it. Salt is needed by our nervous system to function properly, but too much of it may cause high blood pressure.

Salt is a hard, white, **crystalline** solid with a very high melting (801°C) and boiling point (1413°C). It can be made by burning sodium in chlorine gas. Both these substances are very reactive and you would certainly never think of eating either of them yet the substance which is formed when they react is one which your body needs.

$$sodium \; + \; chlorine \; \rightarrow \; sodium \; chloride$$

Sweat is mainly salt solution. Top athletes usually take salt tablets to replace that lost when they sweat

You can see that the properties of the product of the reaction – sodium chloride – are quite different to those of the sodium and chlorine with which we started. An understanding of how the atoms of sodium and chlorine join together (bond) can help to explain this.

In Topic 3.9 we said that inert gases are stable or unreactive because they have full electron energy levels. When elements react to form compounds they do so in a way which enables them to obtain full electron energy levels – so that they form stable compounds. This idea is the basis of chemical bonding. The word bond is used to describe the forces which hold atoms, molecules and ions together.

Freshly cut sodium

Chlorine gas

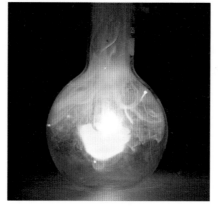

Sodium burning in chlorine gas

The product is salt (sodium chloride)

Ionic bonding

Ionic bonds are usually found in compounds formed when metals, such as sodium, combine with non-metals, such as chlorine. Sodium chloride, formed when they react, is an ionic compound.

A sodium atom has just one electron in its outer energy level ($_{11}Na$ 2,8,1). A chlorine atom has seven electrons in its outer energy level ($_{17}Cl$ 2,8,7).

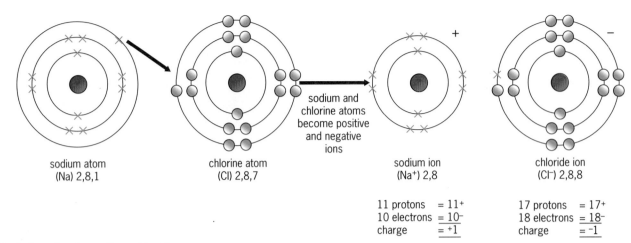

sodium atom
(Na) 2,8,1

chlorine atom
(Cl) 2,8,7

sodium and chlorine atoms become positive and negative ions

sodium ion
(Na⁺) 2,8

chloride ion
(Cl⁻) 2,8,8

11 protons = 11⁺
10 electrons = 10⁻
charge = +1

17 protons = 17⁺
18 electrons = 18⁻
charge = −1

Ionic bonding in sodium chloride

When they react, the outer electron of each sodium atom is transferred to the outer energy level of the chlorine atom. When the sodium atom loses an electron, it becomes a positively charged ion (cation). The chlorine atom, on the other hand, gains that electron becoming a negatively charged ion (anion).

When this type of bond is formed, electrons are transferred from the metal atoms to the non-metal atoms during the chemical reaction. In this way both atoms obtain full outer energy levels and become like the nearest inert gases with a stable structure. The sodium atom has become a sodium ion with an electron configuration like neon, whilst the chlorine atom has become a chloride ion with an electron structure like argon.

Usually only the outer electrons are important in bonding, so we can simplify the diagrams by missing out the inner energy levels.

These oppositely charged ions attract each other very strongly and are pulled, or bonded, to one another by the strong electrostatic forces of attraction between them. These electrostatic forces act in all directions. This gives rise to a three dimensional arrangement of the ions which is known as a lattice. This type of bonding is called ionic bonding. Substances formed by this type of bonding have high melting and boiling points since a lot of energy is required to separate the ions. You will learn more about the properties of ionic substances in the next topic.

Calcium chloride is another important ionic substance. It is used as the antifreeze substance put into concrete mixtures and prevents some of the water in the mixture forming ice crystals. This stops the concrete cracking in cold weather.

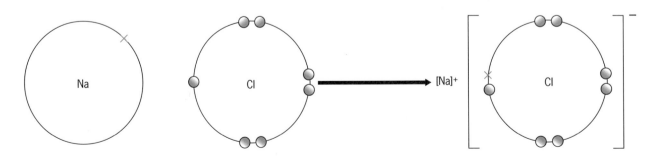

sodium chloride (NaCl)

Simplified diagram of ionic bonding in sodium chloride

The diagram below shows the electron transfers that take place during the formation of calcium chloride. Notice that calcium has two electrons in its outer energy level and chlorine has 7. To obtain full outer energy levels, the calcium must lose two electrons, whereas a chlorine atom only needs to gain one. When these two elements react, the calcium atom gives each of *two* chlorine atoms one electron.

In this case, the compound calcium chloride is formed containing two chloride ions (Cl^-) bound to each calcium ion (Ca^{2+}). Its formula is $CaCl_2$.

This concrete contains calcium chloride

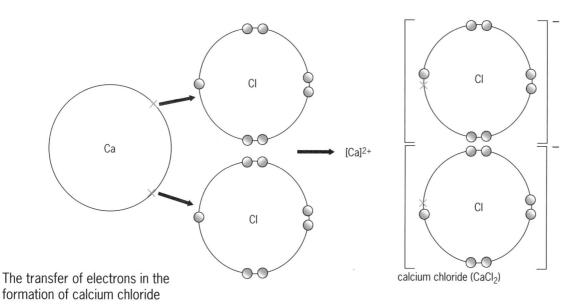

$[Ca]^{2+}$

calcium chloride ($CaCl_2$)

The transfer of electrons in the formation of calcium chloride

---

★ THINGS TO DO

1 Explain what you understand by the following terms:
   a) positive (+) ion
   b) negative (−) ion
   c) ionic bond
   d) electrostatic force of attraction.

2 Using the table on page 111 write down the names of:
   a) three atoms which would form an ion with a charge of +2;
   b) three atoms which would form an ion with a charge of +1;
   c) three atoms which would form an ion with a charge of −1;
   d) 2 metals and 2 non-metals which would form ions with the same size of charge.

3 Use the electron structures shown on page 111 to help you draw diagrams to represent the bonding in each of the following ionic compounds:
   a) calcium oxide (CaO);
   b) magnesium chloride ($MgCl_2$);
   c) lithium fluoride (LiF);
   d) potassium chloride (KCl).

X-raying substances

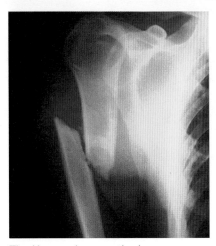

The X-rays show up the bone structure of your arm

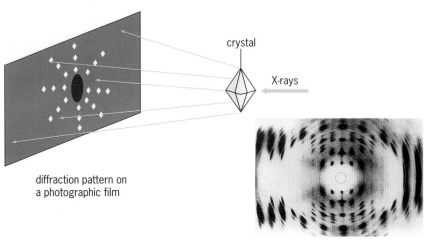

X-rays are used to show the arrangement of the ions in substances like sodium chloride. This technique is called X-ray diffraction

crystal

X-rays

diffraction pattern on a photographic film

Ionic substances are solids at room temperature and have high melting and boiling points. The ions are packed together in a regular arrangement called a lattice. Within the lattice, oppositely charged ions attract one another strongly. How can we find out how the ions are arranged in the **crystal lattice**?

If you have a broken bone in your arm we can get a picture of the break by passing **X-rays** through your arm onto photographic film. The X-rays pass easily through the soft tissues but are blocked by the bone. As a result, the bone appears lighter than the softer tissue surrounding it on the X-ray.

In a similar way X-rays can be used to show the way in which the ions are arranged in crystal lattices. The technique used is called **X-ray diffraction**. The illustration shows the structure of sodium chloride as determined by this technique.

The diagram on page 117 shows only a tiny part of a crystal of sodium chloride. Many millions of sodium ions and chloride ions would be arranged in this way in a crystal of sodium chloride to make up the **giant ionic structure**. Each sodium ion in the lattice is surrounded by six chloride ions and each chloride ion is surrounded by six sodium ions. The forces, or bonds, which hold each of them in place must therefore be very strong.

Not all ionic substances form the same structures. Caesium chloride (CsCl), forms a different structure due to the larger size of the caesium ion compared with that of the sodium ion (see diagram on next page).

Both sodium and caesium chlorides are typical ionic compounds and as such have similar physical properties. Others include rare gems such as sapphire and ruby.

These properties are very much general properties shown by all ionic substances. These properties are listed below.

Ionic compounds:
- are solids at room temperature, with high melting points. This is due to the strong electrostatic forces of attraction holding the crystal lattice together.
- are hard substances.
- cannot conduct electricity when solid because the ions are not free to move.
- dissolve in water. This is because water molecules are able to bond with both the positive and negative ions, which breaks up the lattice and, hence, keeps the ions apart.
- conduct electricity when in the molten state or in aqueous solution. The forces of attraction between the ions are broken and the ions are free to move. This allows an electric current to be passed through molten or aqueous sodium chloride. (See Topic 3.15, page 126.)

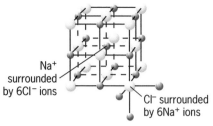

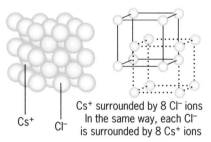

Na⁺ surrounded by 6Cl⁻ ions

Cl⁻ surrounded by 6Na⁺ ions

The arrangement of the ions in sodium chloride

Cs⁺   Cl⁻

Cs⁺ surrounded by 8 Cl⁻ ions
In the same way, each Cl⁻ is surrounded by 8 Cs⁺ ions

The arrangement of ions in caesium chloride – it is different to that of sodium chloride

Some common valencies

| | Valency | | |
	1	2	3
Metals	Lithium (Li⁺) Sodium (Na⁺) Potassium (K⁺) Silver (Ag⁺) Copper (Cu⁺)	Magnesium (Mg²⁺) Calcium (Ca²⁺) Copper (Cu²⁺) Zinc (Zn²⁺) Iron (Fe²⁺) Lead (Pb²⁺) Barium (Ba²⁺)	Aluminium (Al³⁺) Iron (Fe³⁺)
Non-metals	Fluoride (F⁻) Chloride (Cl⁻) Bromide (Br⁻) Hydrogen (H⁺)	Oxide (O²⁻) Sulphide (S²⁻)	
Groups of atoms	Hydroxide (OH⁻) Nitrate (NO₃⁻) Ammonium (NH₄⁺) Hydrogen carbonate (HCO₃⁻)	Carbonate (CO₃²⁻) Sulphate(SO₄²⁻)	Phosphate(V) (PO₄³⁻)

Formulae of ionic substances

Ionic compounds contain positive and negative ions whose charges balance. For example, sodium chloride contains one Na^+ ion to every one Cl^- ion. This idea can be used to write down formulae which show the ratio of the number of ions present in any ionic compound.

The formula of magnesium chloride is $MgCl_2$. This formula is derived from the fact that each Mg^{2+} ion combines with two Cl^- ions. The overall charges balance.

The charge on an ion is a measure of its **valency** or combining power. Na^+ has a valency of 1, but Mg^{2+} has a valency of 2. Na^+ therefore bonds (combines) with only one Cl^- ion, whereas Mg^{2+} bonds with two Cl^- ions so that the overall charges on each ion are equal.

Some elements, such as copper and iron, possess two ions with different valencies.

Copper can form the Cu^+ and the Cu^{2+} ions. It can therefore form two different compounds with chlorine, $CuCl$, copper(I) chloride, and $CuCl_2$, copper(II) chloride. Iron forms the Fe^{2+} and Fe^{3+} ions.

The table above shows the valencies of some of the ions you meet most regularly.

You will notice that the table includes groups of atoms which have charges. For example, the nitrate ion is a single unit composed of one nitrogen atom which has combined with three oxygen atoms, and has one single negative charge. The formula, therefore, of magnesium nitrate would be $Mg(NO_3)_2$. You will notice that the NO_3 has been placed in brackets, with a 2 outside the bracket. This indicates that there are two nitrate ions present for every magnesium ion. The ratio of the atoms present in $Mg(NO_3)_2$ is therefore: 1 Mg : 2N : 6O

★ **THINGS TO DO**

1 Using the information in the table of valencies write the formulae of:
   a) zinc chloride;   b) potassium sulphate;
   c) iron(III) chloride;   d) silver oxide.

2 Using the answers to question **1**, write down the ratio of atoms present for each compound.

3 Calcium oxide is formed when Ca^{2+} and O^{2-} ions combine. Sodium chloride is formed when Na^+ and Cl^- ions combine. The melting points of these two substances are: NaCl 801 °C; CaO 2614 °C. Explain why the melting points of the two substances are so different.

Covalent bonds

Sugar is a very important part of our diet. Many foods contain sugar in one form or another. During respiration, cells in the body convert sugars to carbon dioxide and water, releasing the energy we need.

The formula for sugar is $C_{12}H_{22}O_{11}$. As you will see from its formula sugar contains only atoms of non-metal elements and is a fairly large molecule.

Sugar is a white crystalline solid with a low melting point (180°C). Unlike ionic substances such as sodium chloride (salt), it does not conduct electricity when in aqueous solution (dissolved in water) or when melted.

The reason is that sugar is made up of individual molecules and does not contain ions. Without ions it cannot conduct electricity. Because there are no strong electrostatic forces of attraction present (as there are with ions) it has a much lower melting point than ionic substances.

Since there are no ions in sugar the atoms cannot be held by ionic bonds. They must form a different type of bond so that they gain the stability of the nearest inert gas. They do this by sharing the electrons in their outer energy levels. This type of bond occurs between non-metal atoms, and the bond formed is called a covalent bond. The simplest example of this type of bonding can be seen in the hydrogen molecule, H_2.

Each hydrogen atom in the molecule has only one electron. To obtain a full outer energy level and gain the electron structure of the nearest inert gas – helium – each of the hydrogen atoms must have two electrons. To do this, two hydrogen atoms move close to one another and allow their outer energy levels to overlap as shown.

A molecule of hydrogen is therefore formed with two hydrogen atoms sharing a pair of electrons. This shared pair of electrons is known as a **single covalent bond** and can be represented by a single line.

You may be surprised to find that Corn Flakes contain sugar

Other covalent compounds

There are many other substances which contain non-metals bonded in this way. They include methane, ammonia, and water.

The methane gas burning in this Bunsen burner is a covalent substance

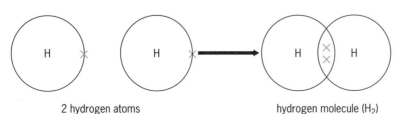

The bonding in a hydrogen molecule

2 hydrogen atoms   hydrogen molecule (H_2)

A model of a hydrogen molecule

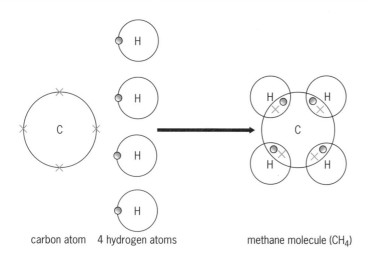

carbon atom   4 hydrogen atoms   methane molecule (CH₄)

The formation of methane

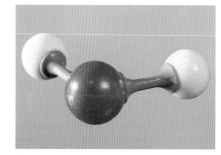

A model of a methane molecule

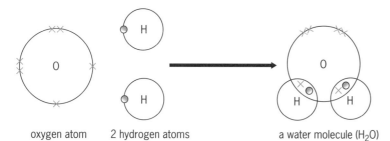

oxygen atom   2 hydrogen atoms   a water molecule (H₂O)

The formation of water

A model of a water molecule

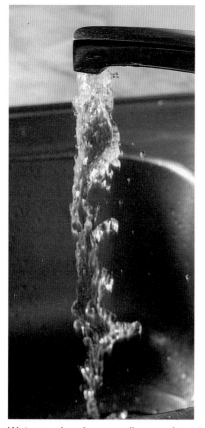

Water molecules are all around us

★ **THINGS TO DO**

1 With the aid of an example in each case explain the following terms:
a) molecule; **b)** single covalent bond.

2 Draw diagrams to represent each of the following covalent compounds:
a) ammonia (NH_3); **b)** oxygen gas (O_2);
c) hydrogen chloride (HCl).

3 The diagram shows the arrangement of the electrons in a molecule of ethanoic acid.
a) Which elements make up this compound?
b) How many atoms are there in the molecule?
c) Between which two atoms is there a double covalent bond?
d) How many single covalent bonds does each carbon atom have?

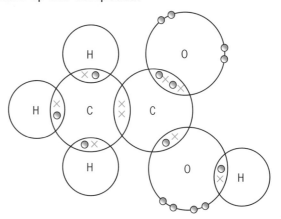

3.13 Giants

The atoms in the substances below are all held together by covalent bonds, yet seem to have quite different properties.

Covalent substances contain molecules whose structures can be classified as either simple molecular or giant molecular.

Simple molecular structures are formed from only a few atoms. They have strong covalent bonds between the atoms within a molecule (**intramolecular bonds**) but have weak bonds between the molecules (intermolecular bonds). One type of weak (intermolecular) bond is a van der Waals' bond (or force). The bigger the molecule the bigger the force between adjacent molecules.

Examples of these simple molecules include solids such as iodine, liquids such as water and gases such as methane (see previous topic).

These substances with simple molecular structures have relatively low melting and boiling points because the forces between the molecules are weak and do not conduct electricity because there are no ions present.

Giant molecular structures have hundreds of thousands of atoms joined by strong covalent bonds. Each part is one enormous molecule sometimes called a **macromolecule**. Sand (silicon(IV) oxide), diamond and graphite all have giant molecular structures. The different structures also help to explain why they have such varied properties.

Graphite and diamond are called **allotropes** because they are made up of the same element – carbon – and exist in the same physical state. However, the difference in their structure helps explain why diamond is the hardest naturally occurring material known, whereas graphite, made up from the same carbon atoms, is one of the softest materials.

In the diamond 'molecule', each carbon atom is held by four others in a regular tetrahedral arrangement (see diagram). This makes the structure very difficult to break – 4 bonds would have to be broken to separate one carbon atom from the others.

The 'molecules' in graphite are organised in layers. Within these layers, each carbon atom is joined to three others by strong covalent bonds,

These substances contain covalent bonds

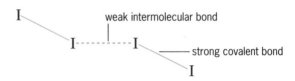

Iodine is a simple molecule with weak intermolecular forces

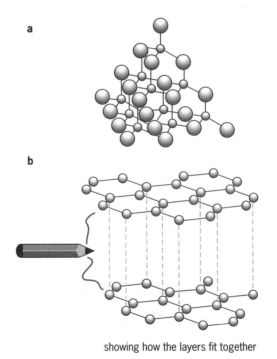

showing how the layers fit together

a) The sand (silicon(IV) oxide) and diamond structures are similar to the one shown. **b)** Graphite has this structure forming a pattern of hexagonal rings. The carbon atoms, like diamond are difficult to separate from one another *but* the forces between the layers are weak. The layers can therefore slide across one another easily, rather like sheets of paper (see diagram on page 120). The graphite is therefore soft and slippery. When you draw layers of graphite are removed from the pencil lead and stick to the paper.

Plastics

Plastics such as polythene are a tangled mass of very long molecules in which the atoms are joined together by strong covalent bonds to form long chains called polymers. Molten plastics can be made into **fibres** by being forced through hundreds of tiny holes in a spinneret as shown in the photograph.

There are two different types of polymer, depending on what type of monomer is used to make them.

Thermoplastics, such as polythene, can be softened when heated but harden again when they are cooled. These plastics can be moulded into any shape.

Thermoplastics contain long polymer chains which can slide around each other.

Within the molecule the atoms are held by strong covalent forces, but the intermolecular forces between adjacent polymer chains are very weak. Their shape can therefore be changed easily.

Other plastics, such as those used to make pan handles and electrical plugs must not soften when they are heated.

These are **thermosetting plastics**. When thermosetting plastics are made, strong covalent bonds form between the chains – the chains become **cross-linked**.

The forces between adjacent chains are much stronger. This makes this type of polymer hard and rigid, even when hot. It also means that thermosetting plastics cannot be softened and so they cannot be remoulded.

Nylon fibres being made

Some uses of thermoplastics

There are no cross-links between the polymer chains of thermoplastics

In thermoplastics, the polymer chains are not linked to each other

Thermosetting plastics have cross-links between the polymer chains

In thermosetting plastics, there are cross-links between the polymer chains

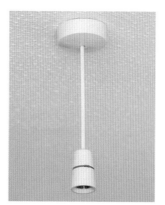
Some uses of thermosetting plastics

Pass the rubber

The rubber that is collected from trees is a runny, white sticky liquid called **latex**. Latex consists of long polymer chains. The eraser rubber you use to correct mistakes is soft, flexible and wears away quickly – you can see the flakes of rubber left on the paper after you have rubbed something out. The rubber used for car tyres is still flexible, but is much harder and does not wear as quickly (although some racing car tyres are made from soft rubber so they grip the ground better).

Eraser rubber is soft and wears away quickly

Collecting latex from a tree

Soft rubber is used for some racing car tyres to give extra grip

Each type of rubber is made from latex. To get different properties, the structure of the latex is altered by adding atoms of a different material – sulphur. The process is called **vulcanising**. The sulphur atoms form strong covalent bonds (attractive forces) between the latex molecules. The more sulphur that is added, the harder the rubber becomes.

Here you can see that by changing the forces which hold the long chains of molecules together we change the properties of the material. The structure affects the way it behaves.

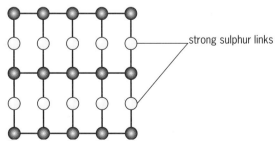

strong sulphur links

The long polymer chains have very weak forces (bonds) between them. The sulphur atoms form much stronger bonds between the chains, making the rubber much harder

Properties of covalent compounds

Just as ionic compounds have similar physical properties, so the two groups of covalent substances described on page 120 (those with simple molecular and those with giant molecular structures) also have similar properties. But these properties are quite different to those of ionic substances.

Covalent compounds:

- as simple molecular substances are usually gases, liquids or solids with low melting and boiling points. The melting points are low because the intermolecular forces of attraction are weak. Giant molecular substances have higher melting points because the whole structure is held together by strong covalent bonds within the giant molecule.

- generally do not conduct electricity when molten or dissolved in water. This is because they do not contain ions. Some molecules, however, react with water to form ions. These substances conduct electricity when in solution. For instance, hydrogen chloride gas produces aqueous hydrogen ions and chloride ions when it dissolves in water:

$$HCl_{(g)} \xrightarrow{\text{water}} H^+_{(aq)} + Cl^-_{(aq)}$$

- generally they do not dissolve in water. However water is an excellent solvent and can interact with and dissolve some covalent molecules better than others.

★ THINGS TO DO

1 Make a table which summarises the properties of the different types of substances you have met in this topic and Topic 3.11. Your table should include examples of ionic substances and covalent substances (simple and giant).

2 With the aid of an example in each case, explain the following words and phrases:
   a) intermolecular bond;
   b) allotrope;
   c) macromolecule;
   d) thermoplastic;
   e) thermosetting plastic;
   f) vulcanising.

3 It is said that both butter and margarine contain simple molecules. What properties do these two foods have which suggest that this statement is correct?

4 Many modern paints are polymers. Inside a sealed tin the polymer chains are rather like those of a thermoplastic – there are no cross-links between the polymer chains. When the paint is used, cross-links form between the chains as the paint hardens.
   a) Using pictures, show the difference between the polymer chains in fresh paint and dried paint.
   b) What do you think causes the polymer chains to form cross-links when applied to wood?

5 Goodyear supplies tyres for Formula 1 racing cars. At times they use a soft compound of rubber which grips the track well. At other times they use a harder compound which does not grip quite so well but lasts longer.
   a) What is likely to be the main difference between the two types of tyre?
   b) Why does the harder tyre not wear as quickly as the softer tyre?
   c) Under what conditions are soft compound tyres and harder compound tyres likely to be chosen?

Metals

Copper is a good conductor of electricity

Magnesium alloys are light and strong

Metals are amongst the most used, and most useful materials on Earth.

The different properties of metals are due to the type of chemical bond between the atoms in the metal, called the **metallic bond**.

Bonding in metals

Metals are also giant structures. The atoms in metals obtain a more stable electron structure by forming another type of bond – a metallic bond. The electrons in the outer energy level of metal atoms can move freely throughout the structure forming a mobile 'sea' of electrons. When the metal atoms lose these electrons they become positive ions. Metals therefore consist of positive ions surrounded by moving electrons.

The negatively charged electrons attract all the positive metal ions and bond them together with strong electrostatic forces of attraction as a single unit. This is the metallic bond. It is this type of bond which gives metals their special properties.

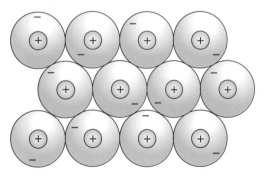

Metals consist of positive ions surrounded by a mobile 'sea' of electrons

The metallic bond and properties of metals

Metals have the following properties (see also Topic 1.1, pages 3–4).

- They have high melting and boiling points due to the strong attraction between the positive metal ions and the mobile 'sea' of electrons.
- They conduct electricity due to the mobile 'sea' of electrons within the metal structure. When a metal is connected in a circuit, the 'free' electrons move towards the positive terminal. At the same time electrons are fed into the other end of the metal from the negative terminal.

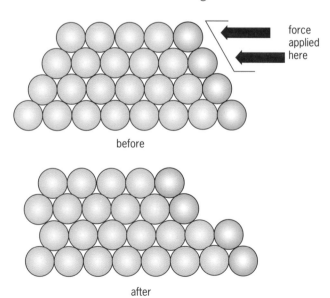

force applied here

before

after

The position of the positive ions in a metal before and after a force is applied

- They are malleable (easily deformed) and ductile (easily stretched). Unlike those in diamond the bonds are not rigid but are still strong. If a force is applied to a metal, rows of ions can slide or slip over one another. They reposition themselves and the strong bonds reform as shown in the diagram on page 124.

    Malleable means that metals can be hammered into different shapes. Ductile means that the metals can be pulled out into thin wires.

- Metals have high densities. This is because the atoms are very closely packed in a regular manner as can be seen in the diagram of the crystal lattice of a metal.

Metals are malleable. A blacksmith hammers metals into the shape he wants

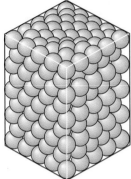

The arrangement of the ions in the crystal lattice of a metal

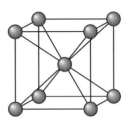
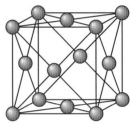

close-packed hexagonal structure (e.g. magnesium, density = 1.74 g cm^{-3})

body-centred cubic structure (e.g. iron, density = 7.87 g cm^{-3})

face-centred cubic structure (e.g. copper, density = 8.92 g cm^{-3})

The ions are arranged like this in different metals and so the metals have different densities

Different metals show different types of packing and in doing so they produce the arrangement of ions shown in the diagram above right.

Alloys

Alloys are metals which contain more than one element (see Topic 1.2, page 6). Steel, for example, is an alloy containing carbon and iron. Bronze is an alloy of copper and tin. Alloys are generally harder and stronger than the separate metals from which they are made. They are made by adding atoms of an alloy element which are of a different size. These act as a barrier and prevent the layers slipping over one another.

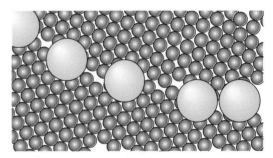

When a second metal is added (making an alloy) the 'layer' atoms of the second metal make it difficult for the edges of the crystals to slip over one another

---

★ THINGS TO DO

1 Explain the terms:
   a) malleable;   b) ductile.

2 Explain why metals are able to conduct heat and electricity.

3 Explain why the melting point of magnesium (649°C) is much higher than the melting point of sodium (97.9°C).

4 Which metal would you use to make each of the following items? Explain your choice in each case:
   a) cooking foil;
   b) electrical wiring;
   c) aircraft construction.

The chlor-alkali industry

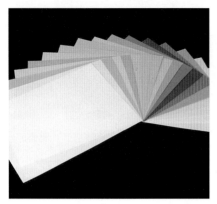

What have all these items got in common?

Everything in the photographs was manufactured using substances obtained from salt. Salt is one of the most important raw materials. It is found in the sea and as underground deposits left by ancient seas as they were left landlocked and evaporated.

Salt is an ionic compound of sodium and chlorine – sodium chloride. Ionic substances are electrolytes – they conduct electricity under certain conditions. During this process the constituent elements are split up when electricity passes through molten salt or an aqueous solution. The process is called electrolysis (see page 26).

The electrolysis of saturated sodium chloride solution (brine) is the basis of a major industry known as the **chlor-alkali industry**. In Britain this process is carried out in Cheshire at a large plant owned by ICI. The salt is obtained from nearby salt beds. These salt beds were formed about 200 million years ago when an ancient sea dried up, leaving thick layers of salt crystals.

The three most important substances produced by the electrolysis of brine are chlorine, sodium hydroxide and hydrogen.

The process is very expensive needing vast amounts of electricity, but is economically worthwhile because all three products have such a range of uses.

Modern methods of electrolysing brine use a **membrane cell**, which produces purer product, causes less pollution and is cheaper to run than other methods.

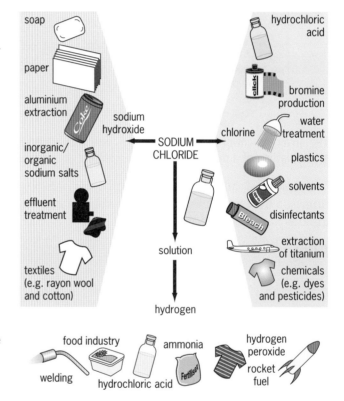

The uses of substances from the chlor-alkali industry

The brine is first purified to remove calcium, strontium and magnesium compounds by a process of **ion exchange**.

The membrane cell is used continuously as fresh brine flowing into the cell replaces that which is used as the products are formed. The cell has been designed to ensure that the products do not mix since chlorine reacts with sodium hydroxide forming sodium hypochlorite (found in household bleach).

The ions in this concentrated sodium chloride solution are:

From the water: $H^+(aq)$, $OH^-(aq)$
From the sodium chloride: $Na^+(aq)$, $Cl^-(aq)$

When the current flows, the chloride ions, $Cl^-(aq)$, are attracted to the anode (the positively charged electrode). Chlorine gas is formed:

$$\text{chloride ions} \xrightarrow{\text{oxidation}} \text{chlorine molecules + electrons}$$
$$2Cl^-(aq) \longrightarrow Cl_2(g) + 2e^-$$

This leaves a high concentration of sodium ions, $Na^+(aq)$ around the anode. The hydrogen ions, $H^+(aq)$, are attracted to the cathode (the negatively charged electrode) and hydrogen gas is produced.

$$\text{hydrogen ions + electrons} \xrightarrow{\text{reduction}} \text{hydrogen molecules}$$
$$2H^+(aq) + 2e^- \longrightarrow H_2(g)$$

This leaves a high concentration of hydroxide ions, $OH^-(aq)$, around the cathode. The sodium ions, $Na^+(aq)$, are drawn through the membrane where they combine with the $OH^-(aq)$ ions to form sodium hydroxide, NaOH, solution. The overall reaction taking place in this process is:

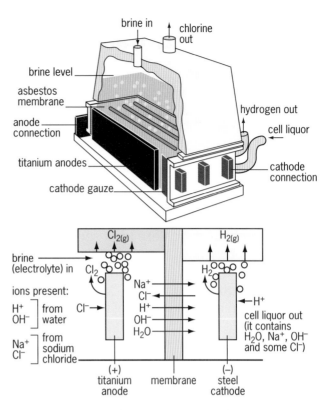

The membrane cell for the electrolysis of sodium chloride

$$\text{sodium chloride} + \text{water} \rightarrow \text{sodium hydroxide} + \text{chlorine} + \text{hydrogen}$$
$$2NaCl(aq) + 2H_2O(l) \rightarrow 2NaOH(aq) + Cl_2(g) + H_2(g)$$

★ THINGS TO DO

1 Sea water contains large quantities of salt. Why is this not used as a source of sodium chloride for the chlor-alkali industry?

2 Why is it important to remove compounds of calcium and magnesium from solutions entering the membrane cell?

3 Account for the following observations which were made when brine, to which a little universal indicator had been added, was electrolysed in the laboratory.
a) The universal indicator initially turns red in the region of the anode but as the electrolysis proceeds loses its colour.
b) The universal indicator turns blue in the region of the cathode.

4 The table opposite shows the % use of sodium hydroxide obtained from the membrane cell.

a) Draw both a bar chart and pie chart using this data.
b) Imagine you are the managing director of a firm making sodium hydroxide. You have to forecast the likely changes in the usage of this chemical over a five year period. What changes do you think may happen in the next five years in the way sodium hydroxide is used? Add reasons for your answer.

Use	%
Manufacture of chemicals	29
Rayon and acetate fibres	16
Soap and detergents	4
Pulp and paper	5
Aluminium oxide	1
Neutralisation	5
Miscellaneous	40

Organising the elements

Go into any supermarket or music store and you will find that things are organised in a way that makes them easy to find. In a music store, you may find the CDs, tapes or discs arranged according to whether they are jazz, classical, or pop. In a supermarket, the different types of food are also organised – cereals are together, dairy foods are gathered together somewhere else, detergents and other household goods are kept separate from foodstuffs.

Similar items are grouped together

The periodic table – devised in 1869 by a Russian Professor of chemistry called Dmitri Mendeléev – is a way in which we can organise each of the chemical elements. The periodic table is organised in such a way that elements with similar properties are grouped together, rather like the foods in the supermarket.

Mendeléev's original periodic table (based on the only 60 elements which were known at the time) has been modified in the light of work carried out by Rutherford and Moseley. Discoveries about sub-atomic particles led them to realise that the elements should be arranged by proton number. This number is shown at the bottom left of the symbols shown for the elements in the periodic table below and on page 200.

Dmitri Mendeléev (1834–1907)

The modern periodic table

1 (I)	2 (II)												3 (III)	4 (IV)	5 (V)	6 (VI)	7 (VII)	0
					1 H hydrogen													4 He helium
7 Li lithium	9 Be beryllium												11 B boron	12 C carbon	14 N nitrogen	16 O oxygen	19 F fluorine	20 Ne neon
23 Na sodium	24 Mg magnesium												27 Al aluminium	28 Si silicon	31 P phosphorus	32 S sulphur	35.5 Cl chlorine	40 Ar argon
39 K potassium	40 Ca calcium	45 Sc scandium	48 Ti titanium	51 V vanadium	52 Cr chromium	55 Mn manganese	56 Fe iron	59 Co cobalt	59 Ni nickel	63.5 Cu copper	65 Zn zinc		70 Ga gallium	73 Ge germanium	75 As arsenic	79 Se selenium	80 Br bromine	84 Kr krypton
85 Rb rubidium	88 Sr strontium	89 Y yttrium	91 Zr zirconium	93 Nb niobium	96 Mo molybdenum	99 Tc technetium	101 Ru ruthenium	103 Rh rhodium	106 Pd palladium	108 Ag silver	112 Cd cadmium		115 In indium	119 Sn tin	122 Sb antimony	128 Te tellurium	127 I iodine	131 Xe xenon
133 Cs caesium	137 Ba barium	89	178.5 Hf hafnium	181 Ta tantalum	184 W tungsten	186 Re rhenium	190 Os osmium	192 Ir iridium	195 Pt platinum	197 Au gold	201 Hg mercury		204 Tl thallium	207 Pb lead	209 Bi bismuth	209 Po polonium	210 At astatine	222 Rn radon
233 Fr francium	226 Ra radium		261 Unq unnilquadium	262 Unp unnilpentium	263 Unh unnilhexium	262 Uns unnilseptium	Uno unniloctium	Une unnilennium	110 *	111 *	112 *							

Atomic numbers (proton numbers): H 1, He 2; Li 3, Be 4, B 5, C 6, N 7, O 8, F 9, Ne 10; Na 11, Mg 12, Al 13, Si 14, P 15, S 16, Cl 17, Ar 18; K 19, Ca 20, Sc 21, Ti 22, V 23, Cr 24, Mn 25, Fe 26, Co 27, Ni 28, Cu 29, Zn 30, Ga 31, Ge 32, As 33, Se 34, Br 35, Kr 36; Rb 37, Sr 38, Y 39, Zr 40, Nb 41, Mo 42, Tc 43, Ru 44, Rh 45, Pd 46, Ag 47, Cd 48, In 49, Sn 50, Sb 51, Te 52, I 53, Xe 54; Cs 55, Ba 56, Hf 72, Ta 73, W 74, Re 75, Os 76, Ir 77, Pt 78, Au 79, Hg 80, Tl 81, Pb 82, Bi 83, Po 84, At 85, Rn 86; Fr 87, Ra 88, Unq 104, Unp 105, Unh 106, Uns 107, Uno 108, Une 109.

* no names yet

- reactive metals
- transition metals
- poor metals
- metalloids
- non-metals
- noble gases

139 La lanthanum 57	140 Ce cerium 58	141 Pr praseodymium 59	144 Nd neodymium 60	147 Pm promethium 61	150 Sm samarium 62	152 Eu europium 63	157 Gd gadolinium 64	159 Tb terbium 65	162 Dy dysprosium 66	165 Ho holmium 67	167 Er erbium 68	169 Tm thulium 69	173 Yb ytterbium 70	175 Lu lutetium 71
227 Ac actinium 89	232 Th thorium 90	231 Pa protactinium 91	238 U uranium 92	237 Np neptunium 93	244 Pu plutonium 94	243 Am americium 95	247 Cm curium 96	247 Bk berkelium 97	251 Cf californium 98	252 Es einsteinium 99	257 Fm fermium 100	258 Md mendelevium 101	259 No nobelium 102	260 Lw lawrencium 103

You can see that, in the modern periodic table, the 112 known elements are arranged in order of increasing proton number.

The key features of the periodic table are:

- Elements with similar chemical properties are found in the same columns or **groups**.
- There are eight groups of elements numbered from 1 to 7. The final group – group 0 – is found in the final column of the table.
- Some of the groups have been given names:

  Group 1: The **alkali metals**
  Group 2: The **alkaline earth metals**
  Group 7: The **halogens**
  Group 0: The **inert** or noble gases

- Between groups 2 and 3 is a block of elements known as the **transition elements**.
- The horizontal rows are called periods and these are numbered 1–7 going down the periodic table.

The periodic table can be divided into two as shown by the bold line. The elements to the left of this line are metals and those on the right are non-metals. The elements which lie on this dividing line are known as **metalloids**. These elements behave in some ways as metals and in others as non-metals.

This metalloid silicon has many uses including silicon chips

The elements with a higher proton number than uranium do not exist naturally but have been made in laboratories. They are all radioactive to some extent, but may be unstable – they change rapidly giving off one or more types of radiation.

★ THINGS TO DO

1 Write down the symbols for the following elements. Also state whether they are 'metals' or 'non-metals' and give their physical state at room temperature.
  a) Vanadium
  b) Krypton
  c) Iridium
  d) Iodine
  e) Barium

2 Give the names and symbols for:
  a) the elements in group 4;
  b) five elements in period 3;
  c) six transition elements;
  d) three metalloids;
  e) eight elements which are radioactive;
  f) the noble gases;
  g) the alkali metals.

3 Use your research skills to find out which of:
  a) the alkaline earth metals is used in flares.
  b) the inert gases is used to fill light bulbs.

c) the group 3 elements is used to make drinks cans.
d) the transition elements is mixed with copper to produce our silver coloured coins.
e) the alkaline earth metals is a major element in bones.

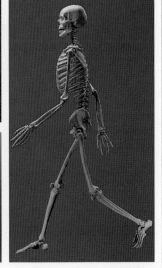

Electron structures and the periodic table

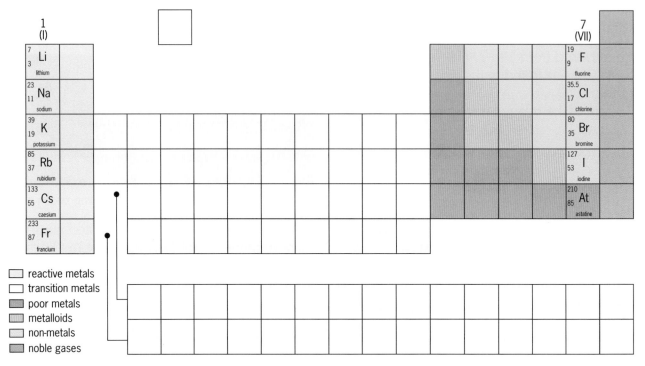

Groups 1 and 7 of the periodic table

The tables show that the number of electrons in the outer energy level of an atom of any element is the same as the number of the group in which it is found in the periodic table. The elements shown in the first table, for example, have one electron in their outer energy level and they are all found in group 1. Along with the elements of group 2, they make up two groups of very reactive metals.

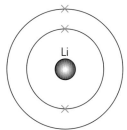

lithium 2,1

Energy level diagram for $_3$Li

Elements found in group 1 of the periodic table

Element	Symbol	Proton number	Number of electrons	Electron structure
Lithium	Li	3	3	2,1
Sodium	Na	11	11	2,8,1
Potassium	K	19	19	2,8,8,1

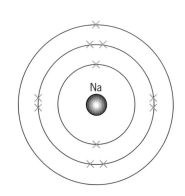

sodium 2,8,1

Energy level diagram for $_{11}$Na

The elements shown in the second table have seven electrons in the outer energy level. They are all reactive non-metals found in group 7.

Elements found in group 7 of the periodic table

Element	Symbol	Proton number	Number of electrons	Electron structure
Fluorine	F	9	9	2,7
Chlorine	Cl	17	17	2,8,7

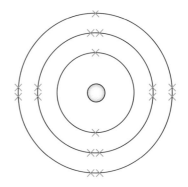

chlorine 2,8,7

Energy level diagram for $_{17}$Cl

The properties of both groups 1 and 7 are dealt with in detail in Topics 3.18 and 3.19.

The elements in group 0, however, are an exception to this rule, as they have 2 or 8 electrons in their outer energy level. This means that their outer energy levels are effectively 'full' and so they are inert – they do not react with other substances to form compounds (see page 136).

Hydrogen is often placed by itself in the periodic table. This is because the properties of hydrogen are unique. However, there are significant comparisons with other elements (see page 137).

★ THINGS TO DO

1 Find the element strontium on the periodic table.
   a) What is its atomic number?
   b) Which group is it in?
   c) How many electrons would you expect it to contain in its outer energy level?

2 Draw the energy level diagrams for the following elements:
   a) Group 1 – lithium and potassium
   b) Group 7 – fluorine
   c) Group 0 – helium and argon

An energy saving light bulb containing Argon

3 Copy and complete the table opposite using information from the periodic table in the previous topic.

4 Name and give the symbols of the other elements in groups 1 and 7 not shown in the tables in the text.

Element name	Symbol	Proton number	Number of electrons	Electron structure	Group number
		5			
	N				
			16		
				2,8,8,2	

The alkali metals

Sodium vapour street lights

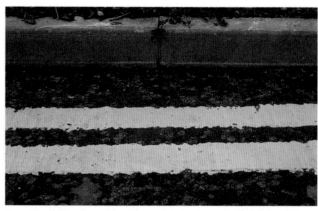

Compounds of potassium are used to make yellow paints for road markings

The photographs show how useful the group 1 elements and their compounds are.

Group 1 consists of the five metals lithium, sodium, potassium, rubidium and caesium and the radioactive element francium. The first three of these, lithium, sodium and potassium, are commonly available for use in school laboratories. They are all very reactive metals, often dangerously so and they are stored under oil to prevent them coming into contact with air or water. These three metals have the following properties.

- They are metals with low densities.
- They are soft metals.
- They are good conductors of heat and electricity.
- When freshly cut with a knife they have shiny surfaces.
- They have relatively low melting and boiling points for metals.
- They burn in oxygen or air with characteristic flame colours, to form white, solid oxides. For example, lithium reacts with oxygen in air to form lithium oxide.

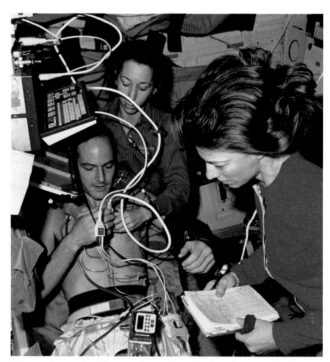

The carbon dioxide produced by the astronauts is absorbed by a lithium compound

$$\text{lithium} + \text{oxygen} \rightarrow \text{lithium oxide}$$
$$4\text{Li(s)} + \text{O}_2\text{(g)} \rightarrow 2\text{Li}_2\text{O(s)}$$

These group 1 oxides all dissolve in water to form alkaline solutions of the metal hydroxide.

$$\text{lithium oxide} + \text{water} \rightarrow \text{lithium hydroxide}$$
$$\text{Li}_2\text{O(s)} + \text{H}_2\text{O(l)} \rightarrow 2\text{LiOH(aq)}$$

Sodium is a soft metal and can easily be cut

- They react very vigorously with water to give an alkaline solution of the metal hydroxide as well as hydrogen gas. For example:

sodium + water → sodium hydroxide + hydrogen
$2Na(s) + 2H_2O(l) → 2NaOH(aq) + H_2(g)$

Potassium is the most reactive of the three elements towards water, followed by sodium and then lithium. Such gradual changes we call trends. Trends are useful to chemists since they allow predictions to be made about the elements we have not seen in action.

Why does reactivity increase as you go down the group?

Considering the group as a whole, the further down the group you go the more reactive the metals become. The table in the previous topic shows the electron structure of the first three elements of group 1. You will notice that in each case the outer energy level contains just one electron. When these elements react they lose this outer electron. By doing this they obtain a full outer energy level. When they lose this electron they become more stable because they obtain the electron structure of a noble gas. You will learn more about the stable nature of these noble gases in Topic 3.20, page 136. For

example, when sodium reacts it loses its outer electron and forms a positive (+) ion (a charged particle).

The sodium ion has the same electron structure as the noble gas neon. This sort of behaviour is seen with all the alkali metals. The smaller the atom is the closer the outer electron is to the nucleus, and the more difficult it is to remove. This is because there is a strong attractive force on it (electrostatic attraction) from the positive protons in the nucleus. As you go down the group, the size of the atoms increases and the outer electron gets further away from the nucleus and therefore becomes easier to remove. This means that as you go down the group the **reactivity** increases.

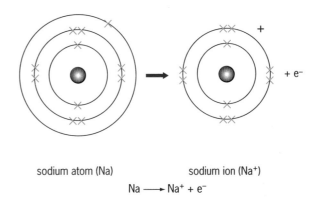

sodium atom (Na)   sodium ion (Na⁺)
$Na → Na^+ + e^-$

A sodium atom becoming a sodium ion

★ THINGS TO DO

1 Use your research skills to find out about as many uses of group 1 metals and their compounds as you can and describe them.

2 Which is the most reactive metal in group 1? Explain your answer.

3 Predict how the group 1 metal caesium will react with:
   a) oxygen;
   b) water.
   In each case write word and balanced chemical equations for the reactions you are describing.

4 Using the following data:
   a) describe the trend shown by boiling and melting points as you go down group 1;
   b) estimate the melting and boiling points of rubidium.

Element	Melting point/°C	Boiling point/°C
Lithium	181	1347
Sodium	98	883
Potassium	64	774
Rubidium	?	?
Caesium	29	679

Halogens

You will have noticed that your local swimming pool always seems to have a special smell. This is because chlorine has been added. Why is this substance put into the water? The chlorine is added to **disinfect** the water. It kills any bacteria or germs which may be harmful to you. Only very small amounts are needed to do this. Chlorine is also added in small quantities to your drinking water to ensure that, when it reaches your home, it is safe for you to drink.

Chlorine is one of the elements found in group 7 (the halogens). Group 7 consists of the four non-metal elements fluorine, chlorine, bromine and iodine, and the radioactive element astatine. Examine the photographs and you will see that fluorine, bromine and iodine are also very useful substances.

Of these five elements, chlorine, bromine and iodine are generally available for use in school laboratories. These elements:

- are coloured as shown in the table.
- exist as diatomic molecules, for example, Cl_2, Br_2 and I_2.
- show a gradual change from a gas (Cl_2) through liquid (Br_2) to solid (I_2)
- form hydrogen halides (e.g. HCl, HBr, HI) which when dissolved in water form an acidic solution.

The chlorine that is added kills the bacteria in the swimming pool water

Fluoride helps the fight against tooth decay

Silver halides based on bromine and iodine are used in photography

Samples of chlorine, bromine and iodine

Halogen	Colour
Chlorine	Pale green
Bromine	Red-brown
Iodine	Purple-black

Colours of some halogens

The reactivity of the halogens

If chlorine is bubbled into a solution of potassium bromide then the less reactive halogen, bromine, is displaced by the more reactive halogen, chlorine. This is called a displacement reaction.

potassium bromide + chlorine → potassium chloride + bromine

$$2KBr_{(aq)} + Cl_{2(g)} \rightarrow 2KCl_{(aq)} + Br_{2(aq)}$$

The order of reactivity of the halogens which has been confirmed by similar displacement reactions, is:

chlorine   bromine   iodine
→
decreasing reactivity

You will notice that the reactivity of these non-metal elements decreases on going down the group. This is unlike the metal elements, of groups 1 and 2. Why does the reactivity decrease as you go down the group? An answer to this question can be obtained by examining the electron structures shown in the table below for chlorine and bromine.

In each case the outer energy level contains 7 electrons. When these elements react they gain one electron per atom and so gain the stable structure of a noble gas. You will learn more about the stable nature of these gases in Topic 3.20, page 136. By doing this each atom becomes a negative (−) ion. For example, when a chlorine atom reacts it gains a single electron and forms the Cl^- ion as shown in the diagram.

The same situation happens with bromine atoms and they become bromide ions (Br^-).

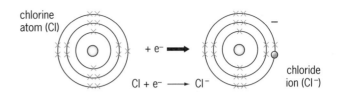

$Cl + e^- \longrightarrow Cl^-$

A chlorine atom forms a chloride ion Cl^-

Chlorine is more reactive than bromine because the incoming electron is being more strongly attracted. This is because the chlorine atom has fewer energy levels and so is smaller. The attractive force on this incoming electron will therefore be greater than in the case of bromine, since the outer energy level of chlorine is closer to the positive nucleus. So as you go down the group the atoms get progressively bigger and they become less reactive.

Element	Symbol	Proton number	Number of electrons	Electron structure
Chlorine	Cl	17	17	2,8,7
Bromine	Br	35	35	2,8,18,7

Electron structure of chlorine and bromine

★ THINGS TO DO

1 Household bleach smells of chlorine because it contains compounds which contain chlorine. These substances are reactive and combine with the dyes in cloth, turning them into colourless compounds. Bleach is probably the most dangerous of household chemicals as shown by the hazard warning label on the bottle in the photograph.

   Look closely at the labels in the photographs and then answer the following questions.

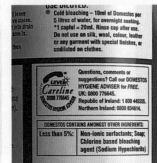

a) Which of the substances shown on the label contains chlorine?
b) What does **irritant** mean?
c) Why should the bottle be stored upright?
d) Why is it important that the bleach ingredients in this bleach are rapidly broken down after use?
e) Bleach is also used to kill germs. Where in your house would you make use of this property?
f) Why is it dangerous to mix bleach with other cleaning products?

2 Write word and balanced chemical equations for the reactions between:
a) bromine and potassium chloride;
b) chlorine and potassium iodide;
c) bromine and potassium iodide;
d) iodine and potassium bromide.
If no reaction will take place, write 'no reaction' and explain why.

The periodic table – the remainder

These lasers contain neon

This airship contains helium

Group 0 – the noble gases

The photographs show situations which involve elements known as the noble gases.

The noble (or inert) gases comprise helium, neon, argon, krypton and xenon as well as the radioactive element radon. This unusual group of gases had not been discovered when Mendeléev published his periodic table. They have only been discovered in the last 100 years, mainly through the work between 1890 and 1898 of the British scientist Sir William Ramsay. They are shown on the extreme right of the periodic table on page 128.

These elements:

- are colourless gases. (However they do produce different colours if an electric current is passed through them. For example, neon is used in the famous lights at Piccadilly Circus.)
- exist as **monatomic molecules**, for example He, Ne, Ar.
- are very unreactive.

No chemical compounds of helium, neon or argon have ever been found or made. However, more recently a number of compounds of krypton and xenon with fluorine and oxygen have been made.

Why are these noble gases so unreactive? They are unreactive because they have electron structures which are stable and very difficult to change. You have seen in Topics 3.18 and 3.19 that they are so stable that other elements attempt to obtain these stable electron structures during chemical reactions.

Although they are unreactive they have many uses some of which are shown in the photographs.

Neon is used in these lights in Hong Kong

The transition elements

In the periodic table on page 128 these elements are found between groups 2 and 3. They are not a proper group because they do not form a downward column in the table. You will be familiar with many of these metals, for example, chromium, iron, nickel, copper and zinc. They are similar in some ways but also very different from the metals you have already become familiar with in groups 1 and 2. Some of these differences include:

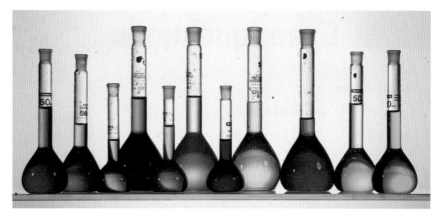

Transition metal compounds are coloured, unlike sodium chloride and magnesium sulphate

- they are harder and stronger.
- they have much higher densities.
- they are much less reactive.
- they form a range of brightly coloured compounds.
- the elements and their compounds often act as catalysts. For example, iron is used in the manufacture of ammonia gas (see Topic 4.13, page 168).

Lithium	Hydrogen	Fluorine
Solid	Gas	Gas
1 electron in outer energy level	1 electron in outer energy level	1 electron short of a full outer energy level
Loses 1 electron to form a noble gas structure	Needs 1 electron to form a noble gas structure	Needs 1 electron to form a noble gas structure
Forms a positive ion	Forms positive and negative ions	Forms a negative ion

Comparing hydrogen with the elements, lithium and fluorine

Hydrogen – the odd one out

Hydrogen is often placed by itself in the periodic table. This is because the properties of hydrogen are unique. However, sensible comparisons can be made with other elements. It is often compared with the elements of groups 1 and 7. But hydrogen cannot fit easily into all the trends shown by either group, as shown in the table.

★ THINGS TO DO

1 Use your research skills to find out about as many uses as possible of:
   a) the noble gases;
   b) the transition elements;
   c) hydrogen.

2 a) Use the periodic table on page 128 to copy and complete the table below.
   b) Look closely at the electron structures for the noble gases and explain why these elements are so stable.

Element	Symbol	Proton number	Number of electrons	Electron structure
Helium				
Neon				
Argon				

3 Why are the transition metals more useful to us than the metals in groups 1 and 2?

Exam questions

denotes higher level questions

1 Use words from this list below to complete the sentences that follow.

gas   liquid   solid

a) A .... takes the shape of a container into which it is poured. (1)
b) When placed in a syringe a .... can be squeezed so that it takes up much less space. (1)
c) A .... has a definite shape. (1)
d) When released into a room a .... will spread out to fill the whole room. (1)

(NEAB, Specimen)

2 a) Complete the table by naming ONE solid, ONE liquid and ONE gas.
   Give a use for each substance named. (6)

	Name	**Use**
Solid		
Liquid		
Gas		

b) The diagram shows the arrangement of particles in a solid and a gas.

solid   liquid   gas

i) Complete the empty box to show how the particles are arranged in a liquid.
ii) Give ONE piece of evidence which clearly shows that there is a lot of empty space between the particles in a gas, but **not** between the particles of a solid. (4)

c) Jim heated a block of ice until it melted to form water. He continued heating until the water boiled. He plotted the change in temperature against time on the graph below.

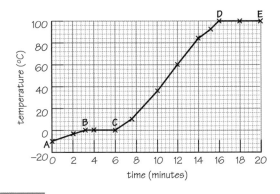

Explain what is happening to the particles at each stage (A–B, B–C, C–D, D–E) as the heating continues. Your answer should include an explanation, in terms of energy, of why there is no temperature change at stages B–C and D–E. (7)

(ULEAC, Specimen)

3 a) Diagram **X** shows the structure of a **solid** metal.

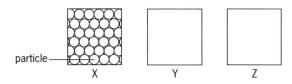

i) Complete diagram Y to show the structure of the metal when it is a **liquid**. (1)
ii) Complete diagram Z to show the structure of the metal when it is a **gas**. (1)
iii) If the metal has a fixed volume but **not** a fixed shape, what state is it in? Is it a gas, liquid or solid? (1)

b) i) When the metal is heated it will **expand**. **Explain** what happens to the metal **particles** as the **solid** metal expands. (2)
ii) When the powdered metal is mixed with sulphur and warmed a chemical reaction takes place. Write down **one** way that you could tell that a chemical reaction has taken place. (1)

(SEG, 1994)

4 Below is a list of elements and their atomic numbers:

hydrogen	1	fluorine	9
helium	2	neon	10
lithium	3	sodium	11
carbon	6	magnesium	12
nitrogen	7	chlorine	17
oxygen	8		

Use the list to answer the following questions.
a) Give the electron arrangement of nitrogen. (1)
b) Name the element with the lowest atomic number and a complete outer shell of electrons. (1)
c) P, Q and R represent three elements in this list. They each combine with hydrogen to form the compounds PH_4, QH_3 and HR. Give the names of P, Q and R. (3)

(WJEC, 1993)

5 The two carbon atoms represented below are isotopes.

ISOTOPE 1.   ISOTOPE 2.

14   ←   mass number   →   12

C   C

6   ←   proton number   →   6

The atoms of both isotopes have six protons.
a) Describe **one other** way in which the isotopes are similar. (1)
b) Describe **two** ways in which they are different. (2)

(NEAB, 1994)

6 a) The element potassium is in the same group of the Periodic Table as sodium. Potassium reacts with chlorine to make potassium chloride. This is sometimes used in cooking instead of common salt.
i) Give the chemical formula of potassium chloride. (1)

ii) By reference to the electronic structures of sodium and potassium, explain why the reaction of sodium with chlorine is similar to the reaction of potassium with chlorine. (1)
b) Fluorine is the most reactive element in group 7 of the Periodic Table. It reacts with all the other elements in the Periodic Table except some of the noble gases. It does not react with helium, argon and neon, but it does react with xenon (Xe) to form xenon fluoride.
i) Explain why fluorine is more reactive than chlorine. (3)
ii) Explain why the noble gases are generally unreactive. (3)
iii) Predict, with reasons, whether radon (Rn) will react with fluorine. (3)

(NEAB, Specimen)

7 This question is about sodium chloride (common salt) which is an important chemical. Sodium chloride can be made by burning sodium in chlorine gas.
a) Use information from the **Data Book** to help to answer these questions.
i) How could you prove by a **chemical test** that a test tube contains chlorine gas? (1)
ii) Why should experiments using chlorine gas be carried out in a fume cupboard? (1)
b) Name an element with similar chemical properties to chlorine. (You might find it helpful to refer to the Periodic Table in the **Data Book**.) (1)

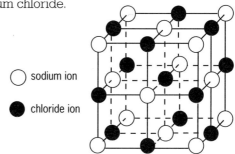

c) Write a word equation for the reaction between sodium and chlorine. (1)
d) Complete the diagrams to show the electronic structure of a sodium atom and a chlorine atom. (2)

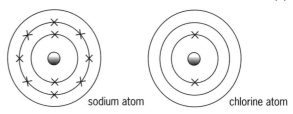

sodium atom   chlorine atom

e) How does a sodium atom change into a sodium ion? (2)
f) The apparatus shown below can be used to electrolyse sodium chloride solution.

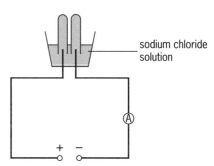

sodium chloride solution

i) Name the product formed at:
    1. the positive electrode
    2. the negative electrode (2)
ii) Give **one** large-scale use of the product formed at the negative electrode. (1)
iii) Give **two** large-scale uses of the product formed at the positive electrode. (2)
iv) The final solution contains Na^+ ions and Cl^- ions.
Name the useful chemical that could be obtained from this solution. (2)

(NEAB, Specimen)

8 Note: The Periodic Table printed on page 128 may help you answer some parts of this question.
a) The diagram below shows the structure of sodium chloride.

○ sodium ion

● chloride ion

i) How does the structure of a sodium **ion** differ from the structure of a sodium **atom**? (1)
ii) How does the structure of a chloride **ion** differ from the structure of a chlorine **atom**? (1)

iii) What type of chemical bond is present in sodium chloride? (1)

b) Use the diagram of sodium chloride to help you to explain why:

i) sodium chloride crystals can be cube shaped; (1)

ii) solid sodium chloride has a high melting point; (2)

iii) solid sodium chloride is an electrical insulator; (2)

iv) molten sodium chloride will undergo electrolysis. (2)

(SEG, 1995)

9 The hydrogen halides (hydrogen fluoride, hydrogen chloride, hydrogen bromide and hydrogen iodide) are important chemicals.
a) The diagram represents a molecule of hydrogen chloride.

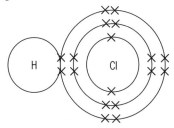

i) What type of particles are represented by the crosses (x)? (1)

ii) What type of chemical bond holds the atoms in this molecule together? (1)

iii) Would you expect hydrogen chloride to be a gas, a liquid or a solid, at room temperature and pressure? Explain your answer. (3)

b) The relative amount of energy required to break the bond in each of the hydrogen halide molecules is shown below.

$$
\begin{array}{ll}
\text{H—F} & 569 \\
\text{H—Cl} & 432 \\
\text{H—Br} & 366 \\
\text{H—I} & 298 \\
\end{array}
$$

One of the important properties of the hydrogen halides is that they dissolve in water to form acids. For example hydrogen chloride reacts with water to form hydrochloric acid.

To form an acid the bond between the hydrogen and the halogen atoms must be broken and ions are formed. The stronger the acid the more molecules that split up to form ions.

i) Which ion must be formed to make a solution acidic? (1)

ii) Which of the hydrogen halides would you expect to react with water to form the strongest acid? Explain your answer. (3)

(NEAB, 1995)

10 The diagram shows an electrolysis cell used in the chlor-alkali industry. Sodium chloride solution is placed in the cell and the products are sodium hydroxide, chlorine and hydrogen.

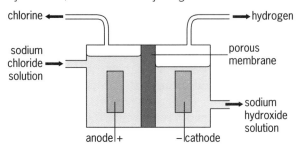

a) i) Write a symbol equation to show what happens at the anode (positive electrode). (2)
ii) Write a symbol equation to show what happens at the cathode (negative electrode). (2)
b) Use your answer to **(a)** to explain how sodium hydroxide solution is made during this electrolysis reaction. (2)

(ULEAC, 1995)

11 Use the Periodic Table on page 128 to help you to answer this question.
a) State **one** similarity and **one** difference in the electronic structure of the elements:
i) across the Period from sodium to argon; (2)
ii) down Group 7 from fluorine to astatine. (2)
b) i) State the trend in reactivity of the Group 1 elements. (1)
ii) Explain this trend in terms of atomic structure. (3)
c) Hydrogen is an element which is difficult to fit into a suitable position in the Periodic Table. Give reasons why hydrogen could be placed in either Group 1 or Group 7. (3)

(NEAB, 1995)

12 Part of the periodic table is shown below.

1 H 1								4 He 2
7 Li 3	9 Be 4	11 B 5	12 C 6	14 N 7	16 O 8	19 F 9	20 Ne 10
23 Na 11	24 Mg 12	27 Al 13	28 Si 14	31 P 15	32 S 16	35 Cl 17	40 Ar 18

Using **only** the elements shown above, give the symbol of:
1 a metal.
2 a non metal.
3 the first element in group 2.
4 an element which consists of molecules which are made up of pairs of atoms.
5 an element which forms a hydroxide which dissolves in water to give an alkaline solution.
6 an element which forms an ion of the type X^-. (6)

(NEAB, Specimen)

4

PATTERNS OF CHEMICAL CHANGE

Chemical reactions

Chemical reactions play an important part in our lives. Chemical reactions in our bodies keep us alive. The process of digestion, for example, is a series of chemical reactions which convert food into chemicals which can be more easily used by our bodies. A reaction inside the cells of our body – respiration – releases the energy we need from sugars in food. (See *GCSE Science Double Award Biology*, Topic 1.9.)

We use chemical reactions in many different ways in and around our homes. At times people may get a burning feeling in their stomach caused by too much acid. By chewing or sucking an antacid tablet, the burning sensation goes away. The acid is neutralised by the chemicals in the tablet.

Generally chemicals are produced to do particular jobs. A few years ago, food stains proved difficult to remove from clothing. Modern 'biological' washing powders contain chemicals called **enzymes** which help to do the job more effectively – they speed up the breakdown of the chemicals in food stains.

Stripping old paint from wood used to be hard work, and sometimes dangerous work if it was burnt away. Chemicals provide a simple solution. This gel, spread onto paintwork, quickly softens the paint so that it can be scraped away easily.

Many industries rely entirely on chemical reactions. In the brewing industry, for example, the reaction between yeast and sugar produces alcohol. Carbon dioxide is another product of the reaction.

Fast or slow

Sometimes we need reactions to occur quickly. The chemicals in fireworks react very quickly with oxygen. Different chemicals are used to produce the different colours. Within a few seconds the reaction is over.

Fireworks – a very fast chemical reaction

The chemical paint stripper softens the paint

Cheese-ripening – a very slow chemical reaction

This glue sets when two chemicals react together

Bronze Age weapons

At other times we prefer reactions to take place slowly. Some cheeses, for example, develop their flavour over several days or even weeks. To slow down the reaction, the cheese is kept in cold conditions while it matures.

For home use, some adhesives consist of two chemicals which must be mixed. The amounts used can make the glue harden slowly, or very quickly.

From the past to the present

Our ancestors discovered that chemical reactions could be used to change naturally occurring rocks (metal ores) in the Earth into more useful substances – metals. Although they did not know about the chemistry of the reactions, they knew how to use them. The metals were used for tools, weapons, and for everyday implements.

Through time, people have learned more and more about the nature of materials – what they contain and how their atoms and molecules are arranged. This knowledge

allows us to solve many problems which face the modern world. Some of these problems are to do with health – some drugs (chemicals) can often provide almost immediate relief from illness and others, taken regularly, can help to prevent asthma attacks (see photo below).

As people find out more and more about chemicals, and the way they react, we can take further steps towards making the world a better place for everyone to live.

Chemicals are used to combat many medical problems

★ THINGS TO DO

1 Explain the following terms:
  a) respiration;
  b) enzymes;
  c) chemical reaction;
  d) antacid.

2 Devise an experiment to show that washing powders containing enzymes speed up the

breakdown of chemicals in food stains. Discuss the details with your teacher and then, if possible, carry out the experiment you have suggested.

3 Make a list of the processes you can find in this topic in which chemical reactions play an important part.

Helping industry

As new techniques have been developed, the processes used in chemical and manufacturing industries have become more complex. Manufacturing industries which use chemical processes are always thinking of ways to improve the yield (how much of the product they can get from the raw materials they use) and the speed at which they can make it (making it faster means that they can keep prices low). Chemists and chemical engineers are always looking for ways to control the rate at which chemical reactions take place. They have discovered that there are five things which affect the rate at which a chemical reaction takes place:

- the surface area of the reactants (the substances you start with);
- the concentration of the reactants in solution (or pressure if gases are involved);
- the temperature at which the reaction is carried out;
- the brightness and colour of light;
- the use of chemicals called catalysts.

Measuring the rate at which a reaction takes place

Some reactions take place slowly. Others take place quickly. The speed at which a reaction takes place is the **rate of reaction**. Many methods can be used to measure the rate of reaction. Basically, however, these methods either:

- measure the rate at which one of the reactants (the starting materials) is used up or
- measure the rate at which one of the products is produced.

If one of the products is a gas then we could measure the amount of gas produced. The gas could be collected in a gas syringe or in an upturned burette or other container (see diagram). The rate could be recorded as the volume of gas produced every ten seconds or every minute, for instance $20\,cm^3/10\,s$.

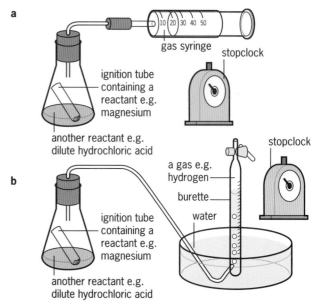

a) Using a gas syringe to measure the gas produced
b) Using an upturned burette to measure the gas produced

Measuring the change in mass as the reaction takes place

Another method used, if a gas is one of the products, is to record how the mass of the reactants changes as the reaction takes place (see diagram). The rate could, for example, then be recorded as the amount of mass lost (in grams) every 10 seconds, for instance $0.3\,g/10\,s$.

If a coloured product is formed in the reaction it is possible to use the formation of this product to act as a guide for measuring the rate of reaction. Reactions which produce a rapid change of colour are sometimes

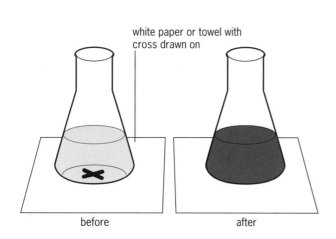

The iodine clock reaction produces a rapid change in colour

Computers can be used to follow the progress of a chemical reaction

known as **clock reactions**. The rate of this type of reaction can be monitored using the techniques shown in the diagram.

In many cases computers and sensors can be used to measure mass, time and even the colour changes which occur.

★ THINGS TO DO

1 A group of pupils tested the effect of acid on some limestone from a nearby quarry. They used two different strengths – 0.25 M and 0.5 M. They put small pieces of limestone in a conical flask and added the acid. They then measured the loss of mass every 4 minutes.

Time /min	Mass lost with dilute acid/g	Mass lost with stronger acid/g
4	3.5	6.9
8	5.9	9.2
16	8.6	10.1
20	9.2	10.2
24	9.6	10.2
28	9.9	10.2
32	10.0	10.2
36	10.0	10.2

a) What would they need to do to make sure their tests were fair?
b) Plot a graph of their results for the dilute acid. Your graph should have the change of mass on the vertical axis and time on the horizontal axis.

c) What information can you get from the graph about how quickly the reaction was occurring at different times?
d) Plot a graph for the concentrated acid on the same axes.
e) Make a list of similarities and differences between the reactions for each acid.
f) What conclusions can they draw from their investigation?
g) Describe another method which could be used to measure the rate of reaction of each acid. Why would using two different methods improve their conclusions?
h) Why do you think the reaction stopped for each concentration when about 10 g had been lost?

2 a) What is meant by the term reaction rate?
b) What conditions would speed up the rate at which food deteriorates?

3 Using the information given in this topic, explain why reaction rates are so important to manufacturing industries.

What happens when chemicals react?

Clints and grikes in limestone paving

Limestone reacting with sulphuric acid

This photograph shows the effect of acid rain on limestone rock in the countryside. The acid reacts with the limestone, dissolving it in places.

The same reaction can be seen in the laboratory if limestone (calcium carbonate) is placed into sulphuric acid.

When this takes place, the calcium carbonate reacts with the sulphuric acid. Calcium sulphate, carbon dioxide gas and water are formed.

$$\text{calcium carbonate} + \text{sulphuric acid} \rightarrow \text{calcium sulphate} + \text{carbon dioxide} + \text{water}$$

$$CaCO_{3(s)} + H_2SO_{4(aq)} \rightarrow CaSO_{4(s)} + CO_{2(g)} + H_2O_{(l)}$$

(the reactants)   (the products)

For the acid to react with the calcium carbonate the particles of each substance must come into contact with each other. This happens because the particles of the liquid are able to move freely. As they do so they collide with the calcium carbonate particles.

When some of the acid particles meet some particles of calcium carbonate they react (if they have sufficient energy). Carbon dioxide gas and water are formed, and a particle of calcium sulphate is left behind.

(Notice that the original calcium carbonate and sulphuric acid no longer exist – they have reacted to form new substances. Carbon

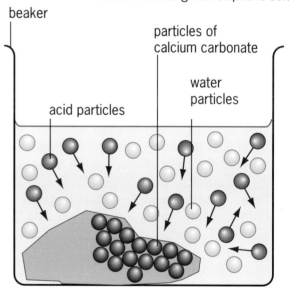

Acid particles collide with the calcium carbonate during the reaction

dioxide gas will therefore rise to the surface – a visible sign that a reaction is taking place.)

In a real situation thousands of particles will be reacting at any time, so lots of gas will be seen. As more and more acid particles react with the calcium carbonate, there are fewer acid particles left in the solution – the acid becomes more and more dilute, or weaker. Because there are fewer acid particles, the rate of reaction slows down. At the same time, the mass of the reactants decreases as carbon dioxide is released into the atmosphere.

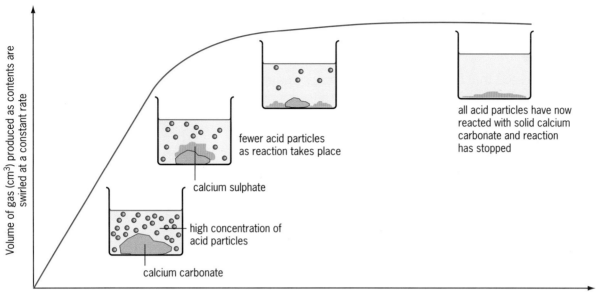

The rate of reaction slows down

Why does the reaction eventually stop?

There are two factors which affect the end of the reaction:

- when all of the acid particles have reacted with calcium carbonate particles the reaction will stop. It is quite likely that there may still be some calcium carbonate left, but because there are now no acid particles, no further reaction can take place.

- when all the calcium carbonate has reacted with the acid. There may still be some acid left, but because there is no calcium carbonate left, no further reaction can take place.

Generally, to avoid waste (leaving unreacted substances) the amounts of the reactants are carefully calculated to ensure that they are fully used.

★ THINGS TO DO

1 Imagine repeating the above experiment using hydrochloric acid and calcium carbonate chips. What do you think would happen to the rate of reaction if several different concentrations of acid were tested? (If the concentration is doubled then the number of acid particles in solution doubles.)
a) Describe your **hypothesis** clearly, using diagrams to help your explanation.
b) Plan some tests which you could do to test your hypothesis. You will need, in particular, to describe any safety precautions which will be needed, and how you will change the concentration of the acid. How can you plan to

avoid the situation where the gas is produced too quickly to be measured?
  When your teacher has approved your plan, carry out your tests and produce a written report.
  Before you finish, is there any way in which you could get back-up data, perhaps using another method, to support your conclusions?

2 Magnesium reacts with dilute hydrochloric acid producing hydrogen gas.
a) Does the magnesium react faster at the start of the reaction or at the finish?
b) Explain your answer to **a)**.

What affects the rate of reaction? (1)

Surface area

In Topic 2.2, we said that limestone (calcium carbonate) could be used to neutralise acid soils. It is often used to counter the effects of acid rain on soils and lakes. Powdered limestone is used because it neutralises the acidity faster than if lumps of limestone are used.

The reaction between acid and limestone (either lumps or powder) can be easily carried out in the laboratory. The photograph shows the reaction between dilute hydrochloric acid and limestone in lump and powdered form.

$$\text{hydrochloric acid} + \text{calcium carbonate} \rightarrow \text{calcium chloride} + \text{carbon dioxide} + \text{water}$$

$$2HCl_{(aq)} + CaCO_{3(s)} \rightarrow CaCl_{2(aq)} + CO_{2(g)} + H_2O_{(l)}$$

The rate at which the reactions occur can be found by measuring either:

- the volume of the carbon dioxide gas which is produced, or
- the decrease in mass of the reaction mixture with time.

These two methods are generally used for reactions in which one of the products is a gas.

The apparatus shown in the photograph is used to measure the decrease in mass of the reaction mixture. The mass of the conical flask and the reaction mixture is measured at regular intervals. The total decrease in mass is calculated from each reading of the balance, and this is plotted against time as shown in the graph.

The reaction is generally at its fastest in the first minute. This is shown by the slopes of the graph – the steeper the slope, the faster the rate of reaction. You can see from the graphs that powdered limestone reacts faster than limestone in lump form.

The surface area has been increased by powdering the limestone.

The powdered limestone reacts faster with the acid than limestone in the form of lumps

The total mass can be measured against time

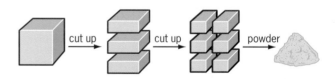

Sample results for the limestone acid experiment using lumps of and powdered limestone

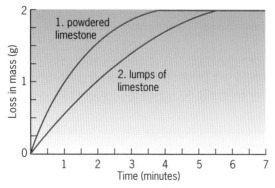

A powder has a larger surface area

The acid particles meet the limestone particles more often, and the reaction takes place quicker. Increasing the surface area of the limestone increases the rate of reaction.

Concentration

When sodium thiosulphate and hydrochloric acid are mixed, a yellow **precipitate** of sulphur is produced. The solution becomes increasingly difficult to see through as more and more sulphur is formed.

$$\begin{array}{c} \text{sodium} \\ \text{thiosulphate} \end{array} + \begin{array}{c} \text{hydrochloric} \\ \text{acid} \end{array} \rightarrow \begin{array}{c} \text{sodium} \\ \text{chloride} \end{array} + \begin{array}{c} \text{sulphur} \\ \text{dioxide} \end{array} + \text{sulphur} + \text{water}$$

$$Na_2S_2O_{3(aq)} + 2HCl_{(aq)} \rightarrow 2NaCl_{(aq)} + SO_{2(g)} + S_{(s)} + H_2O_{(l)}$$

The rate of this reaction can be followed by recording the time taken for a given amount of sulphur to be precipitated. This can be done by placing a conical flask containing the reaction mixture onto a cross on a piece of paper (see photographs).

As the amount of sulphur increases, the cross becomes more and more difficult to see and finally disappears from view. The time taken for this to occur is a measure of the rate of this reaction.

The precipitate of sulphur obscures the cross

If several experiments are carried out using different concentrations, then the rate of reaction for each concentration can be compared by plotting the data from each experiment on the same axes, as shown in the graph.

You can see that when the most concentrated sodium thiosulphate solution was used, the reaction was at its fastest. This is shown by the shortest time taken for the cross to be obscured.

In a more concentrated solution there are more particles so collisions occur more frequently. The more often they collide the faster they react. This means that the rate of a chemical reaction will increase if the concentration of reactants is increased.

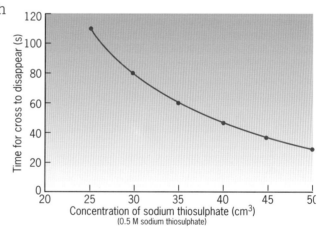

Sample data for the sodium thiosulphate acid experiment at different concentrations of sodium thiosulphate

---

★ THINGS TO DO

1 The following results were obtained, using apparatus similar to that shown on page 148, for the reaction between excess limestone chippings and dilute hydrochloric acid.

Time (min)	0	1	2	3	4	5	6	7	8
Loss in mass (g)	0	1.1	2.0	2.1	2.6	2.8	3.0	3.0	3.0

a) Plot a graph of the results like the one shown on page 148.

b) Which of the results would seem to be incorrect? Explain your answer.
c) How long did the reaction last?
d) Write a word and balanced chemical equation to represent the reaction taking place.
e) Explain why the mass of the conical flask and its contents decreases during the reaction.
f) Explain using particle theory how the rate changes during this reaction.

4.5

What affects the rate of reaction? (2)

Temperature

The reaction between sodium thiosulphate and hydrochloric acid can also be used to study the effect of temperature on the rate of a reaction. The graph opposite shows some sample results of experiments with sodium thiosulphate and hydrochloric acid which have been carried out at different temperatures.

You can see from the graph that the rate of the reaction is fastest at high temperature. When the temperature at which the reaction is carried out is increased, the energy of the particles increases – they move faster and collide more often.

This increases the rate at which the particles of sodium thiosulphate and hydrochloric acid collide, and also, because the particles have more energy, the collisions which occur are more likely to form products. In general, reactions take place faster at higher temperatures.

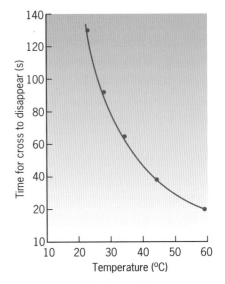

Sample data for the sodium thiosulphate/acid experiment at different temperatures

Light

Some chemical reactions are affected by light. Photosynthesis (see *GCSE Science Double Award Biology*, Topics 2.4–2.6), for example, only takes place when sunlight falls on leaves which contain the green pigment, **chlorophyll**.

Photographs are produced by a reaction between light and silver salts on the film. In this case the silver salts on the film are changed into silver when light falls on them. The more light which falls onto the film, the faster the reaction.

$$Ag^+(s) + e^- \rightarrow Ag(s)$$

A negative is formed as light falling onto the photographic film produces a chemical change. (See illustration.)

When the film is developed, areas which received a lot of light appear darker than those which received less light.

To get good quality photographs the amount of light must be just right to allow the reaction on the film to take place effectively.

Photosynthesis needs sunlight

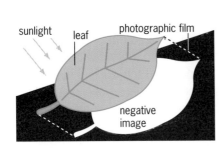

A negative forms as light falls onto the film

These photographs were taken with too much (left) and too little (right) light

★ THINGS TO DO

1 Films are developed by placing them in developing solution. They must be developed for just the right time otherwise they will be ruined. The chart shown here is included with home developing kits and shows the **developing times** under different conditions.

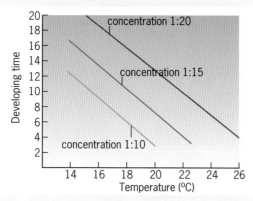

The graph shows developing times as a function of temperature/concentration

a) What is the developing time if the concentration of the developer is 1:15 and the temperature of the solution is 20°C?
b) One photographer started with 60cm^3 of concentrated developer and added 600cm^3 of water. Which of the graphs for concentration should he use to decide on the developing time?
c) Use the information in the graphs to write a sentence which (i) describes the connection between temperature and developing time, and (ii) describes the connection between concentration and developing time. For each one, explain the connection using ideas about particles.
d) A photographer notices that when he started developing his film (in a solution with a concentration of 1:20) the temperature was 20°C. Towards the end of the developing time he noticed that the temperature had fallen to 16°C. What could he do to make up for the change in temperature?

2 Baking powder, which contains sodium hydrogencarbonate, can be used to make bread and cakes. When mixed with water it releases carbon dioxide gas, which helps the dough to rise.

a) Why must the dough mixture be placed in a warm place to rise?
b) What would happen if the temperature was too cold?
c) What would be the effect of using (i) too little (ii) too much baking powder?

3 One of the world's most famous landmarks, the Acropolis in Athens, is now in more danger than it has been for two thousand years.

The problem is caused by acid rain. In the past thirty years the amount of sulphur dioxide which pours into the atmosphere around Athens has increased tremendously. The sulphur dioxide dissolves in rain, forming a solution containing sulphuric acid. The acid then falls on the marble, dissolving it.

a) As the amount of sulphur dioxide in the atmosphere increases, the acid rain becomes more concentrated. What effect do you think this will be likely to have on the marble of the Acropolis? Add reasons for your answer.
b) Plan how you could investigate your ideas, including any safety precautions you will need to take. Carry out your investigation when your teacher has approved your plan.
c) In your account, add a section which gives some suggestions about (i) other factors which might affect the rate at which the marble is damaged, and (ii) how the problems could be overcome.

Catalysts

Canoes are built by placing glass fibre matting over a mould and soaking the fibre in a liquid resin. The resin sets hard. The fibres give added strength.

The same materials can be used to repair damaged canoes. The repair kit contains a small amount of matting, the resin and a compound called a catalyst. The catalyst speeds up the hardening process.

A catalyst is a substance that speeds up a reaction and is left chemically unchanged at the end of the reaction. The enzymes in biological washing powders are catalysts – they speed up the breakdown of biological stains in clothing. Some types of adhesive contain two compounds which must be mixed before use. One contains a catalyst – the more catalyst that is added the faster the adhesive hardens.

Glass fibre repair kit for canoes

Catalytic converters prevent air pollution such as this

Catalytic converters

Car exhaust fumes contain gases such as carbon monoxide (CO) and nitrogen oxides (e.g. NO). Carbon monoxide is poisonous. Nitrogen oxides dissolve in moisture in the air, forming a dilute solution of nitric acid. This is one of the constituents of acid rain.

Inside a normal exhaust system, some of the carbon monoxide and nitrogen oxides are changed into less harmful gases.

carbon monoxide + oxygen → carbon dioxide
$$2CO_{(g)} + O_{2(g)} → 2CO_{2(g)}$$

nitrogen(II) monoxide + carbon monoxide → nitrogen + carbon dioxide
$$2NO_{(g)} + 2CO_{(g)} → N_{2(g)} + 2CO_{2(g)}$$

They are, however, changed very slowly. To increase the rate at which the reactions take place (and so to reduce the amount of these pollutants), European regulations now mean that all new cars have to be fitted with **catalytic converters** as part of their exhaust system (see illustration).

The catalyst in the converter speeds up the reactions, converting the pollutants to carbon dioxide and nitrogen which are naturally

A section through a catalytic converter

present in the air. Catalytic converters, however, can only be used with unleaded petrol. Impurities in petrol also 'poison' the catalyst and it has to be replaced every five or six years.

Speed increases production

In industry it is important to make sure that reactions take place as quickly as possible. By increasing production, costs can be kept low. Over 90% of industrial processes use catalysts for this purpose. During reactions involving catalysts, the catalyst itself does not take part in the reaction – it is left chemically unchanged at the end of a reaction.

Catalysts at work

In the laboratory, the effect of a catalyst can be seen using hydrogen peroxide.

$$\text{hydrogen peroxide} \rightarrow \text{water} + \text{oxygen}$$
$$2H_2O_{2(aq)} \rightarrow 2H_2O_{(l)} + O_{2(g)}$$

Hydrogen peroxide is often used as a bleach – it breaks down naturally forming water and releasing oxygen (which is the bleaching agent). The rate of decomposition at room temperature is very slow. Manganese(IV) oxide, (MnO_2), is a catalyst which can be used to speed up the reaction – oxygen is produced more rapidly.

The rate at which this occurs can be followed by measuring the volume of oxygen gas produced with time and plotting a graph as shown in the diagram.

At the end of the reaction, the manganese(IV) oxide can be filtered off and used again. In reactions involving gases, any dirt or impurities on the surface of the catalyst reduce how well it works. It is said to have been 'poisoned'.

In general, the points to remember about catalysts are:

- a small amount of catalyst produces a significant change in the rate of reaction.
- catalysts are not changed, chemically, during a reaction, but can change physically. During the decomposition of hydrogen peroxide described above, the manganese (IV) oxide powder which is left behind is much finer than at the start of the reaction.
- different reactions need different catalysts.

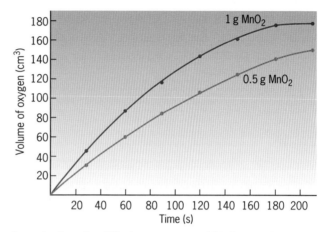

Sample data for differing amounts of MnO_2 catalyst

★ THINGS TO DO

1 Two experiments were performed to find whether the amount of manganese(IV) oxide added to hydrogen peroxide affected the rate at which the reaction took place. The pupils measured the volume of gas produced at different times.
a) Plot graphs of volume of gas against time using their results.
b) What effect does increasing the amount of manganese(IV) oxide have on the rate at which oxygen is produced? Use their results to explain your answer.
c) Why do the slopes of the graphs become less steep during the later stages of the reaction?
d) What volume of gas was produced by 0.3g of manganese(IV) oxide after 50 seconds?

e) How long did it take for 60cm³ of gas to be produced when the experiment was carried out using 0.5g of the manganese(IV) oxide?
f) Is the data reliable enough to say that the rate of reaction increases as more and more catalyst is added? If not, suggest what could be done to obtain more reliable information.

2 Why do some people consider catalytic converters not to be as environmentally friendly as suggested in their advertising material?

3 Explain the following statement: Industrial processes become more economically viable if a catalyst can be found for the reactions involved.

Time /s	0	30	60	90	120	150	180	210
Vol for 0.3 g / cm³	0	29	55	79	98	118	133	146
Vol for 0.5 g / cm³	0	45	84	118	145	162	174	182

Well brewed

Some reactions are brought about by living things; a process called **biotechnology**. **Yeast**, for example, is a fungus – a living organism. You may have seen wild yeasts as a white powder on grapes and plums.

Yeast is one of the ingredients in kits for making beer and wine at home. When the ingredients are mixed, the yeast changes the sugar into alcohol. This reaction is called **fermentation**. Under the right conditions, the yeast cells multiply rapidly, speeding up the reaction. Bubbles of carbon dioxide gas can be seen rising from the fermenting solution. The same reaction is used on an industrial scale in the production of beers and wines.

The equation below summarises the reaction.

$$\text{sugar} \xrightarrow{\text{yeast}} \text{alcohol (ethanol)} + \text{carbon dioxide} + \text{energy}$$

$$C_6H_{12}O_6{}_{(aq)} \longrightarrow 2C_2H_5OH_{(l)} + 2CO_2{}_{(g)}$$

The temperature of the reaction must be very carefully controlled. The best temperature for this process is 37°C. The reaction which takes place inside the yeast cells depends on chemicals called enzymes (see *GCSE Science Double Award Biology*, Topic 1.8). The enzymes are biological catalysts. They are **protein** molecules which are damaged when the temperature rises above 50°C. Above this temperature fermentation ceases.

Respiration

Most organisms obtain the energy they need to survive by respiration, during which sugars such as glucose are broken down into carbon dioxide and water. Energy is released during the process. For respiration to occur, oxygen is needed. This is **aerobic respiration**.

$$\text{sugar} + \text{oxygen} \longrightarrow \text{carbon dioxide} + \text{water} + \text{energy}$$

$$C_6H_{12}O_6{}_{(aq)} + 6O_2{}_{(g)} \longrightarrow 6CO_2{}_{(g)} + 6H_2O_{(l)}$$

Some organisms, such as yeast, can break down glucose without oxygen – a process called **anaerobic respiration**. This process also releases energy. During anaerobic respiration, the glucose may be broken down into alcohol and carbon dioxide – this is alcoholic fermentation. This process is used in the baking and brewing industries.

Yeast growing on fruit

Home brewing kit

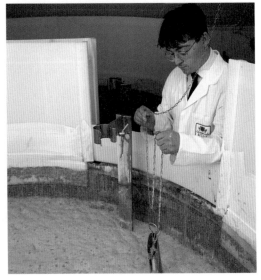
Industrial fermentation

The baker

Many traditional bakers still use yeast and sugar in their bread mixture. The baker is more interested in the carbon dioxide which is released as the sugar ferments. The carbon dioxide makes the dough rise as the bubbles of carbon dioxide are trapped in the mixture. When baked, the bread has a light, 'fluffy' appearance.

In many parts of the world people use unleavened bread. This is made without yeast and does not rise in the same way.

The spaces are caused by trapped carbon dioxide gas

From the grape to the bottle

Ripe grapes contain sugar. When picked at just the right time, they are covered with a layer of wild yeast. After the grapes have been picked, they are crushed. The juice from the grapes is collected in huge vats. The yeast cells multiply rapidly (they double in number about every three days) under carefully controlled conditions, causing the sugar in the juice to change to **alcohol**. The end product is wine – another product of fermentation.

★ THINGS TO DO

1 Fermentation needs to take place fairly quickly. The manufacturers of a home brew kit must suggest the best temperature on their instruction sheet.
   a) How do you think the rate of fermentation depends on the temperature of the fermenting solution? Add reasons supporting your prediction.
   b) Plan your own test to check your ideas. You can use a sugar solution to which yeast can be added. If your teacher approves your plans, carry out your test.

   Write a clear report describing what conclusions can be drawn from your results. Add a section evaluating your investigation and the data.

2 The graphs show what happens to yeast under different conditions.

a) What do you think are the ideal conditions for yeast to grow?
b) What is happening to the yeast population (the number of yeast cells) in graph 1?
c) At what temperature does the yeast begin to reproduce? At what stage of the bread making process do you think this happens?
d) At what temperature do you think the yeast cells die? When does this happen in the bread making process?
e) What is being produced as the yeast reproduces and grows?

3 Use your research skills to find out more about the problem of alcohol abuse. Why is alcohol abuse a problem in our society?

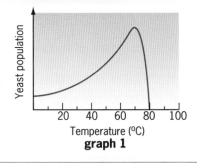

graph 1

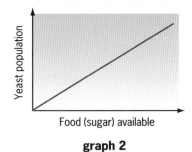

Food (sugar) available
graph 2

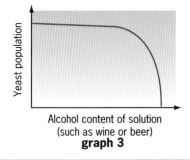

Alcohol content of solution (such as wine or beer)
graph 3

Mouldy cheese

Making cheese and yoghurt

Living organisms are responsible for the useful changes which are used to make cheese and yoghurt. In both cases **microbes** are used to break down sugars (especially lactose) in curdled milk.

To make cheese, lactic acid-producing bacteria are added to milk. These bacteria cause the milk to become acidic. At just the right stage, rennet is added to the 'milk'. A white, solid curd forms very quickly. The solid curd is separated from the water (whey) and is then strained to allow excess water to drain away. The curd is then placed into moulds and microbes (either bacteria or fungi) are then used to break down the curd, softening it and giving it the characteristic 'cheesy' flavour. This process is known as ripening. Unlike alcoholic fermentation, the particular microbes used in cheese and yoghurt making are decay microbes. Instead of breaking down the sugar into alcohol, they form lactic acid.

Some of the microbes are found naturally in the cheese. Others are added to create a different texture or flavour.

Mouldy food

In the examples we have seen so far, microbes are used in a useful way to produce things which we need. Microbes are also responsible for decay. After a while, most foods are affected by **micro-organisms** – fungi or bacteria. Because these are living organisms they need oxygen, water and food to survive. In warm conditions, where these are present, they multiply rapidly.

Prepacked foods must be sold before they 'go off'. Most are now stamped with a 'Use by' or 'Best before' date – after this date they will soon reach the stage where they become harmful to health.

Milk contains a particular type of sugar compound called lactose. Lactose breaks down in water, releasing energy and forming lactic acid. Certain enzymes produced by bacteria speed up the breakdown of the lactose into lactic acid. This gives the milk a sour taste (because it is becoming more acidic as more and more lactic acid is produced).

When cheese decays, it becomes soft. Moulds often grow over the surface, eventually penetrating deep below the surface.

The process is called **spoilage**. Spoilage may be due to:

- chemicals in the food reacting with oxygen in the air. The chemicals are oxidised, forming new chemicals which can affect the flavour and appearance of the food.

The holes in cheeses are formed by gases released by microbes and the mould formed by microbes shows in blue cheeses

Oxidation causes the browning of apples when exposed to the air. Moulds grow rapidly in warm, damp conditions

- organisms (bacteria or fungi) called spoilage organisms. These organisms must change the food before they can begin digesting it. They do that by producing enzymes. The enzymes break down the chemicals in the food, some of which produce acids giving the food its characteristic sour taste.

Preventing decay

There are many ways to slow down the rate at which food decays but most rely on slowing down the activity of the enzymes and killing the microbes.

The rate at which enzymes break down food depends on their temperature. In warm, moist conditions microbes multiply rapidly, producing enzymes. Enzymes work quickly in warm conditions and so break down food quickly. The lower the temperature, the slower the activity of the enzymes. At 4°C – the temperature of the average home refrigerator – enzyme activity is very slow, and so spoilage is slowed down.

At −18°C – the temperature of industrial freezer units – enzyme activity is almost stopped, so food can be stored for very long periods without damage.

Chemicals called preservatives may be added to the food. Sulphur dioxide, for example, is often used as a preservative in soft drinks

Preventing oxidation. Many foods are vacuum packed (packed in plastic packaging from which all air has been removed) to prevent oxidation of the chemicals

Increasing the acidity or alkalinity of the food. Preservatives such as jams and pickles have high levels of acidity which slow down the rate at which enzymes work

Using high concentrations of salt or sugar, for example in jams and salted fish. Preservation by this method relies on the salt or sugar drawing water from the enzymes, preventing them being involved in the decay process

Heating to high temperatures. Blanching (boiling followed by rapid cooling) for example, raises the temperature of fruit and vegetables to a point at which enzyme activity is stopped. If the foods are then frozen, they will remain in good condition for several months

Irradiating the food. Exposing the food to ionising radiation kills any microbes which may be present

The most common way of preserving food is to refrigerate or freeze it. This may be done even with foods preserved in some of the ways described above. The lower the temperature, the slower the decay

The many ways of slowing down the rate at which food decays

★ THINGS TO DO

1 Until 1850 people thought that the micro-organisms which caused food to decay were formed by the food itself. Louis Pasteur carried out an investigation which showed that this idea was wrong, and that the micro-organisms which caused decay came from the air and the surroundings.

   Write a plan describing how you think Louis Pasteur carried out his investigation. You should say clearly what you think he did, and what observations he would have made.

2 A further method of preventing decay is drying or dehydration. Explain how drying a foodstuff can prevent its decay.

3 Spoilage or decay requires certain agents such as bacteria or moisture. Name four other agents of a biological or physical nature which help the process of decay.

Cool reactions

When reactions take place energy is transferred. When fuels burn, for example, energy is transferred in the form of heat.

$$\text{methane (natural gas)} + \text{oxygen} \rightarrow \text{carbon dioxide} + \text{water} + \text{heat energy}$$

$$CH_4(g) + 2O_2(g) \rightarrow CO_2(g) + 2H_2O(l) + \text{energy}$$

Some fuels release energy quickly, rapidly warming the surroundings. Others burn more slowly, releasing energy slowly. They do not have the same 'warming' effect.

A similar reaction takes place in the cells of the body – respiration. During respiration sugars react with oxygen, releasing carbon dioxide, water and energy – the energy which keeps our bodies warm.

$$\text{glucose} + \text{oxygen} \rightarrow \text{carbon dioxide} + \text{water} + \text{energy}$$

$$C_6H_{12}O_6(aq) + 6O_2(g) \rightarrow 6CO_2(g) + 6H_2O(l) + \text{energy}$$

Under normal conditions the cells release just enough energy to meet our needs. At other times, such as when we exercise, we need energy faster. Respiration takes place faster – releasing energy faster – and so the warmer we get.

Reactions such as these which release energy (which may be transferred to the surroundings or may simply raise the temperature of the reactants) are exothermic reactions.

Getting started

Although combustion (the burning of substances in oxygen) is an exothermic reaction, some energy is needed initially to get the reaction started. Fireworks, for example, must be lit (using the blue touchpaper) before they will begin to burn. The energy needed to start the reaction (provided by the burning blue touchpaper) is called the **activation energy**.

Like all other chemical reactions, combustion takes place faster at higher temperatures. Once the chemicals in the firework are ignited the reaction temperature is increased and they burn faster, releasing enough energy to keep them burning.

The reactions inside fireworks are examples of exothermic reactions which release energy rapidly.

Energy is needed to start any reaction. In some cases, such as photosynthesis and rusting, little activation energy is needed and the reactions can start at relatively low temperatures. Other reactions need to be heated strongly before they begin – they have a higher activation energy.

Peat burns slowly, releasing energy slowly

Coal releases energy quickly

The burning blue touchpaper provides the energy to make the firework burn

Cooling down

Sporting people regularly suffer injuries such as sprains. If the injury is cooled there is less chance of swelling and bruising. Ice packs solve the problem – but they do not contain ice. Most contain chemicals which dissolve quickly in water. As they do they draw energy from the water, cooling it quickly.

If the ice pack is applied to an injury, it cools the area around the injury. This type of reaction, which is accompanied by a fall in temperature (or which absorbs energy from the surroundings) is an **endothermic reaction**.

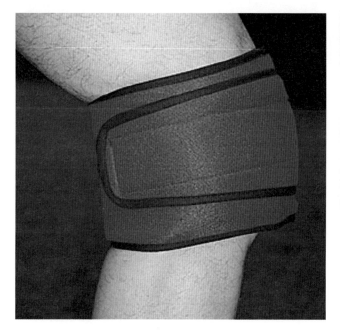

The ice pack (*above* and *left*) reduces swelling and bruising

★ THINGS TO DO

1 Compounds such as potassium nitrate, sodium nitrate or potassium chloride can all be used in the type of ice pack described above.

Make a list of factors which could affect the temperature fall during dissolving. Plan how you could test your ideas. Be sure to consider how to avoid any hazards which may be present. When your teacher approves your plan, carry out your tests.

Prepare a report which could be used by the manufacturers of the ice packs. You should say how they could get the best cooling effect in their product.

2 Samantha said that when sherbet dissolved in her mouth it felt cold. She said, 'The reaction between sherbet and water is endothermic.'

Sherbet is a mixture of citric acid and sodium hydrogencarbonate.
a) How could you carry out a laboratory test to find out whether the reaction was endothermic?
b) If the reaction is endothermic, it must be carefully controlled. If heat is drawn from the tongue too quickly it causes an injury similar to a burn. Write a detailed plan describing how you could find out which of the ingredients has the greatest endothermic effect.

These athletes are powered by combustion

We obtain most of our energy from the combustion of fuels, such as hydrocarbons and from the combustion of foods (respiration).

When, for example, natural gas burns in a good supply of air it produces a large amount of energy.

$$\text{methane} + \text{oxygen} \rightarrow \frac{\text{carbon}}{\text{dioxide}} + \text{water} + \frac{\text{heat}}{\text{energy}}$$

$$CH_4(g) + 2O_2(g) \rightarrow CO_2(g) + 2H_2O(l) + \frac{\text{heat}}{\text{energy}}$$

During this process, the **complete combustion** of methane, energy is transferred to the surroundings as heat. It is an exothermic reaction. If only a limited supply of air is available then the

reaction is not as exothermic (less energy is released in the same time) and a poisonous gas, carbon monoxide, is produced.

$$\text{methane} + \text{oxygen} \rightarrow \frac{\text{carbon}}{\text{monoxide}} + \text{water} + \frac{\text{heat}}{\text{energy}}$$

$$2CH_4(g) + 3O_2(g) \rightarrow 2CO(g) + 4H_2O(l) + \frac{\text{heat}}{\text{energy}}$$

This process is known as the **incomplete combustion** of methane.

The energy change which takes place during a chemical reaction can be shown by an energy level diagram. The illustration shows the energy level diagram for the complete combustion of methane which is exothermic.

Energy level diagram for the complete combustion of methane

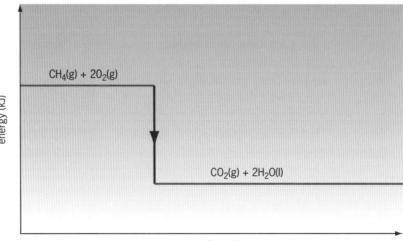

$CH_4(g) + 2O_2(g)$

$CO_2(g) + 2H_2O(l)$

energy (kJ)

progress of reaction

When any chemical reaction occurs the chemical bonds in the reactants have to be broken – this requires energy. When the new bonds in the products are formed energy is given out, as shown in the illustration.

The **bond energy** is defined as the amount of energy associated with the breaking or making of one mole of chemical bonds in a molecular element or compound. You will need to turn to Topic 4.15 to find out about moles.

The energy needed to break, or make, bonds between different elements can be found in tables, such as that shown.

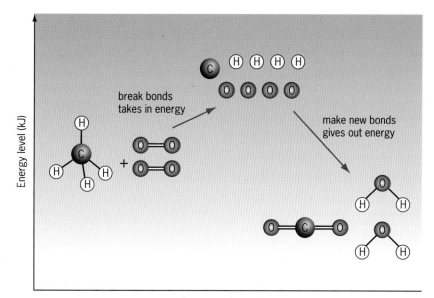

The breaking and formation of bonds during the combustion of methane

Bond energy data

Bond	Bond energy (kJ mol^{-1})
C—H	435
O=O	497
C=O	803
H—O	464

For the reaction shown above – the combustion of methane – we can use the data in the table to calculate how much energy is transferred at each stage.

The stages in the reaction

When the methane burns, it combines with oxygen to form carbon dioxide and water. As the reaction proceeds:

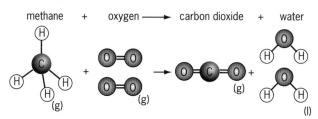

- the bonds between the atoms of hydrogen and carbon (in the methane) are broken;
- the bonds between the oxygen atoms in the oxygen molecules are broken;
- new bonds are formed between carbon atoms and oxygen atoms to form carbon dioxide;
- new bonds are formed between hydrogen and oxygen atoms to form water.

We can work out how much energy is needed to break the bonds using the information in the table as follows:

Breaking 4 C—H bonds in methane requires
$4 \times 435 = 1740$ kJ
Breaking 2 O=O bonds in oxygen requires
$2 \times 497 = 994$ kJ

Total energy needed to break the bonds in 1 mole of the reactants = 2734 kJ of energy

Making 2 C=O bonds in carbon dioxide gives out
$2 \times 803 = 1606$ kJ
Making 4 O—H bonds in water gives out
$4 \times 464 = 1856$ kJ

Total energy transferred from reactants as bonds are formed = 3462 kJ of energy

There is therefore more energy released when the new bonds form the products of the reaction than when the bonds between the reactants are broken. Energy will therefore be transferred to the surroundings.

$$\begin{array}{ccc} \text{energy} \\ \text{difference} \end{array} = \begin{array}{c} \text{energy required} \\ \text{to break bonds} \end{array} - \begin{array}{c} \text{energy given out} \\ \text{when bonds} \\ \text{are made} \end{array}$$

$$= 2734 - 3462$$
$$= -728 \text{ kJ}$$

The negative sign shows that the chemicals are losing energy to the surroundings, that is, it is an exothermic reaction. A positive sign would indicate that the chemicals are gaining energy from the surroundings – that the reaction is endothermic.

The energy stored is called the **enthalpy** and given the symbol **H**. The change in energy going from reactants to products is called the **change in enthalpy** and is shown as ΔH (pronounced 'delta H'). ΔH is called the **heat of reaction**. For an exothermic reaction ΔH is negative and for an endothermic reaction ΔH is positive.

When fuels, such as methane, are burned they require energy to start the chemical reaction off. As you saw in the previous topic, this is known as the activation energy, E_A (see graph). In the case of methane reacting with oxygen it is the energy involved in the initial bond breaking (see illustration). The value of the activation energy will vary from fuel to fuel.

Endothermic reactions are much less common than exothermic ones. In this type of reaction energy is absorbed from the surroundings so that the energy of the products is greater than that of the reactants. The reaction between nitrogen and oxygen gases is endothermic, (see graph).

$$\text{nitrogen} + \text{oxygen} \rightarrow \text{nitrogen(II) monoxide}$$
$$N_2(g) + O_2(g) \rightarrow 2NO(g)$$

Dissolving is often an endothermic process. For example, when ammonium nitrate dissolves in water the temperature of the water falls indicating that energy is taken from the surroundings. Photosynthesis and thermal decomposition are other examples of endothermic processes.

In equations it is usual to express the ΔH value in units of kJ mol^{-1}.

For example,

$$CH_4(g) + 2O_2(g) \rightarrow CO_2(g) + 2H_2O(l)$$
$$\Delta H = -728 \text{ kJ mol}^{-1}$$

Energy level diagram for methane + oxygen showing ΔH and E_A

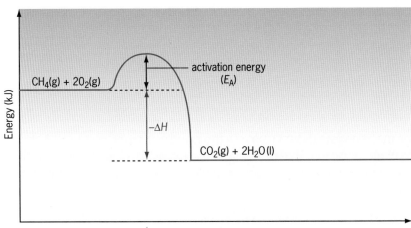

Energy level diagram for nitrogen + oxygen

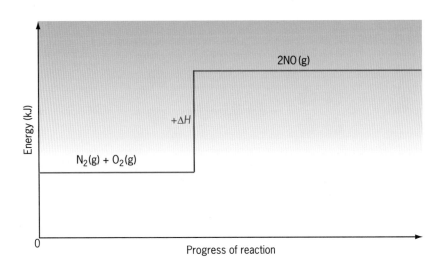

This ΔH value tells us that when one mole of methane is burned in oxygen 728 kJ of energy are released. This value is called the **enthalpy of combustion** of methane (molar heat of combustion of methane).

The effect of catalysts

A catalyst increases the rate by providing an alternative reaction path with a lower activation energy. The activation energy is the energy barrier which reactants must overcome, when their particles collide, to successfully react and form products (see graph).

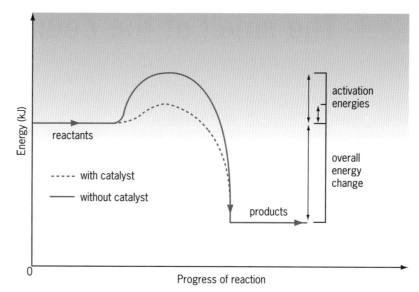

Energy level diagram showing activation energy with and without a catalyst

Biological washing powders contain enzymes (biological catalysts) which can break down the chemicals in food stains such as eggs, coffee and blood

★ **THINGS TO DO**

1 Explain the following statements.
 a) The combustion of magnesium is complete.
 b) The combustion of petrol in a car engine is usually incomplete. (*Note* – petrol contains some molecules of the formula C_8H_{18}.)

2 State which of the following processes is endothermic and which is exothermic.
 a) The forming of a chemical bond.
 b) The breaking of a chemical bond.

3 Using the information given in this topic and Topics 4.15 and 4.18 about the combustion of methane, calculate how much energy is released if:
 a) 0.5 moles of methane are burned?
 b) 5 moles of methane are burned?
 c) 4 g of methane are burned?
 (*Note.* You will find it helpful to refer to Topics 4.15 and 4.18 when you are doing this question.)

4 How do catalysts speed up chemical reactions?

5 Draw fully labelled energy level diagrams for the reactions represented by the following equations.
 a) $HCl_{(aq)} + NaOH_{(aq)} \rightarrow NaCl_{(aq)} + H_2O_{(l)}$
$$\Delta H = -57\,kJ$$
 b) $6CO_{2(g)} + 6H_2O_{(l)} \rightarrow C_6H_{12}O_{6(aq)} + 6O_{2(g)}$
$$\Delta H = +2820\,kJ$$

The smell of the country

To many people the smell of the countryside means farmyard manure. That familiar smell is caused largely by a gas called ammonia. Compounds of ammonia are present in all animal waste. Scattered on the soil, they help to produce better, and bigger crops.

In the first half of the 20th century animal manures were used as natural **fertilisers**, helping to improve the texture of the soil and to replace important nutrients taken from the soil by the previous year's crops. The main nutrients needed by plants are nitrogen, potassium and phosphorus.

The nutrients taken from the soil must be replaced otherwise the crops grown in the following years would become weaker and weaker. Some experiments carried out over a long period at a research station show that a soil which contains the right balance and amount of nutrients can provide five times the amount of food as one which has not been supplied with fertiliser.

As the population increased, there was a greater demand for food. Farmers needed to produce more food from the land. But this meant that more nutrients were lost each year as more crops were grown. There was not enough natural manure to replace the nutrients which were being removed. **Artificial fertilisers** became increasingly used as the only way to replace the nutrients which were needed, although they could not improve the texture of the soil.

Artificial fertilisers

There are many different types of fertiliser. Some contain high proportions of potash. Others contain high proportions of nitrates. Most, however, are compound fertilisers – they contain all three of the major elements in different proportions. The relative amounts of the different elements present are shown on the packaging.

When they dissolve, nitrate, phosphate and potassium ions are made available to the plants via their roots. Each has a part to play in the way the plant develops.

Spreading manure has its disadvantages

Farming on a large scale

Artificial fertilisers

Element	Benefits	Symptoms of shortage
Nitrogen	Promotes 'green' growth	Stunted growth; leaves become yellow due to lack of chlorophyll
Potassium (as potash)	Helps resist disease and cold weather	Leaves become yellow and curve inwards
Phosphorus	Encourages photosynthesis, helps plants use energy efficiently	Plants grow slowly. Seeds and fruit are small

The hidden dangers

When fertilisers are added to soil they dissolve. They may then be washed into rivers and streams and may even end up in our drinking water. The problems may not immediately become apparent – it can take many years for nitrates to be washed through the soil into water. Nitrates can cause particular problems for wildlife and humans.

In water which contains large amounts of nitrates algae multiply rapidly, turning the water green. As they multiply, the nitrate levels in the water fall. When these organisms die they decay and eutrophication takes place. The microbes that cause decay use up oxygen in the water. Eventually there may be insufficient oxygen in the water to support fish and other organisms and they die.

Nitrates in drinking water can cause particular problems for young babies. The nitrates change into compounds called nitrites.

This river has been badly affected by nitrates washed off farmland

When these are absorbed into the blood, they reduce the amount of oxygen which the blood can carry. The baby turns blue – a symptom of 'blue baby syndrome' which must receive immediate medical attention.

★ THINGS TO DO

1 Soil test kits can be used by gardeners to find out whether they need to add fertiliser to their soils, and if so which type of fertiliser is needed.

   Follow the instructions given in a soil test kit to find out whether some samples of soils are lacking any nutrients which are needed by plants.

2 Imagine you are a research scientist. You have been given the task of developing a new product for use as a fertiliser. What properties would this product need to have to make it a suitable material for a farmer to use? (You could get some extra ideas from reading the next Topic.)

A soil test kit

Nitrogen

Nitrogen is vital for growing plants. It encourages growth of the green parts of the plant – the stem and the leaves. Almost 80% of the air is nitrogen but unfortunately the vast majority of plants cannot use nitrogen in this form. They obtain the nitrogen they need from ammonium compounds in the soil. Because the ammonium compounds are soluble, they dissolve in water in the soil. The resulting solution can then be absorbed through the roots of the plant.

Naturally occurring

Some compounds which supply nitrogen are made by the action of bacteria on animal wastes and by the action of lightning on the nitrogen in the air. They replace some of the nitrogen removed from the soil by growing plants.

On land which is intensively farmed, these processes cannot replace the nitrogen which is taken from the soil. To add extra nitrogen farmers often rotate crops – moving them from one field to another from year to year.

Every so often they include a crop of clover or peas. Their roots contain bacteria which are able to absorb nitrogen from the air and convert it into nitrogen compounds.

These bacteria are called **nitrogen-fixing bacteria**. When ploughed back into the soil, the nitrogen compounds are released making them available for other crops. Additional nitrogen, where it is needed, is supplied by artificial fertilisers.

A clover crop is grown to improve the nitrogen content of the soil

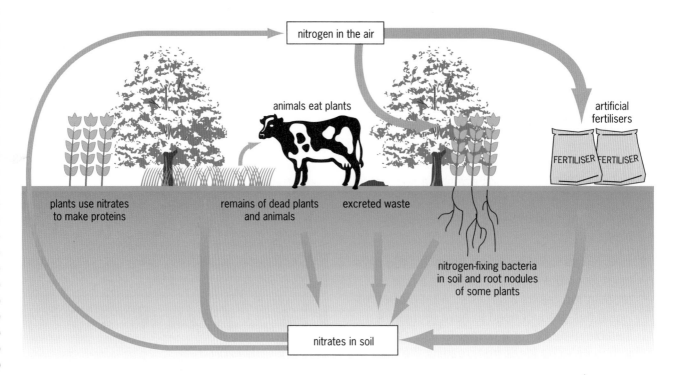

The **nitrogen cycle**

Artificial fertilisers

Fertilisers which have a high nitrogen content are described as **nitrogenous fertilisers**.

Some common nitrogenous fertilisers are:

ammonium nitrate	NH_4NO_3
ammonium phosphate	$(NH_4)_3PO_4$
ammonium sulphate	$(NH_4)_2SO_4$
urea	$CO(NH_2)_2$

You can see from the formula for each of the compounds that they all contain N – the symbol for the element nitrogen.

A nitrogenous fertiliser

Preparing ammonium nitrate

Fertilisers are salts (see page 38) made by reacting a soluble base (alkali) with an acid. As a general rule,

$$acid + base \rightarrow salt + water$$

To make ammonium nitrate, ammonium hydroxide (the base) is mixed with nitric acid.

ammonium hydroxide	+	nitric acid	$\rightarrow$	ammonium nitrate	+	water
$NH_4OH_{(aq)}$	+	$HNO_{3(aq)}$	$\rightarrow$	$NH_4NO_{3(aq)}$	+	$H_2O_{(l)}$

The most economic way of preparing a salt is by using carefully measured amounts of the acid and base which are just sufficient for neutralisation to take place.

The alkaline ammonia solution is neutralised by the acid. Different ammonium salts can be made by using different acids. (Using sulphuric acid, for example, you can make ammonium sulphate.)

heat

1 Measure out 20 cm³ of ammonia solution in a conical flask.
2 Add a few drops of universal indicator.
3 Put 20 cm³ of your acid into a measuring cylinder.
4 Use a dropping pipette to add acid to the ammonia solution until it is just neutralised.
5 Make a note of how much acid was needed.
6 Wash out the conical flask. Now put another 20 cm³ of ammonia solution in your clean flask. Do not use indicator this time, but add just the amount of acid that is needed to neutralise the ammonia. Shake the solution well.

7 Pour your salt solution into an evaporating basin and heat it gently using a bunsen burner so that it just begins to evaporate.
8 When crystals of salt begin to form above the liquid surface, stop heating. Use the tongs to pick up the evaporating basin and pour the contents into a crystallizing dish.
9 Leave for several days until crystallization is complete then weigh the fertiliser you have made.

Preparation of an ammonium salt

★ THINGS TO DO

1 Prepare a sample of an ammonium salt using the method shown. Write an account of your experiment, including diagrams where they will help. Include words such as dissolving, filtering, evaporation, condensation and neutralisation where they are appropriate. Include a word equation describing the reaction which took place.

2 Prepare a flow chart showing the nitrogen cycle in action.

3 Pea plants can be grown by placing them in a jar containing sand (sand does not contain any of the nutrients which plants need). Plan how you could carry out some tests to find out
a) whether fertilisers affect plant growth, and
b) how different types of fertilisers affect the growth.
Carry out your tests when your teacher approves your plan.

Write an article for a gardening magazine describing how your results could help the keen home gardener. You can add other research information if you wish.

The need for nitrogen

Manufacturers need huge amounts of nitrogen to make fertilisers on an industrial scale. The nitrogen is used to prepare ammonia, which is then used to prepare ammonium salts. There are three key stages in the production of ammonium nitrate, as shown in the illustration.

At each stage, chemical principles are applied to make sure that the reactions take place as efficiently as possible. Efficient production methods mean lower costs.

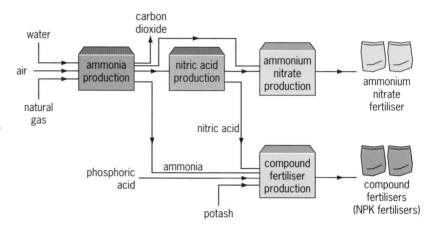

The key stages in the production of ammonium nitrate

Ammonia production

The large scale manufacture of ammonia is carried out using the **Haber process**. Ammonia is a gaseous compound containing the elements nitrogen and hydrogen. The raw materials needed to make it are:

- nitrogen, obtained by cooling air to a point where the nitrogen liquefies followed by fractional distillation;
- hydrogen, made by the reaction between natural gas (methane) and steam.

methane + steam → hydrogen + carbon dioxide
$$CH_4(g) + 2H_2O(g) \rightarrow 4H_2(g) + CO_2(g)$$

Nitrogen is an unreactive gas, and so special conditions (very high temperature and pressure) are needed to make it react with hydrogen to form ammonia.

The reaction between the nitrogen and hydrogen takes place inside a steel pressured vessel which is fitted with an iron catalyst. The catalyst speeds up the reaction between the nitrogen and hydrogen.

nitrogen + hydrogen ⇌ ammonia
$$N_2(g) + 3H_2(g) \rightleftharpoons 2NH_3(g)$$

The equation shows arrows travelling in both directions. This is because some of the ammonia which is produced immediately changes back into hydrogen and nitrogen.

Ammonia plant at Billingham, Cleveland, UK

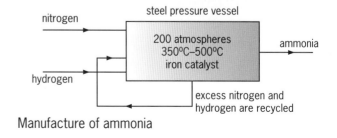

Manufacture of ammonia

The reaction is **reversible** – it can work in both directions. The conditions of high temperature and pressure help make sure that more ammonia is produced than is changed back into hydrogen and nitrogen. The amount of ammonia formed is called the **yield**. If, for example, the yield is 15%, the other 85% of the gases are recycled – returned into the system to react again.

Although most of the ammonia produced is used in the manufacture of fertilisers, some is used for explosives, and for making nitric acid.

Nitric acid production

When the oxides of non-metals dissolve in water they produce acid solutions. We saw on page 82, for example, that when sulphur dioxide dissolves in water it forms sulphuric acid. Nitric acid is produced in a similar way when nitrogen dioxide is dissolved in water.

The first step in the manufacture of nitric acid is one in which the ammonia is oxidised to form nitrogen monoxide and water. The reaction is carried out at high temperatures (800°C) and a catalyst of platinum and rhodium is used to increase the rate of reaction.

ammonia + oxygen $\rightarrow$ nitrogen monoxide + water + energy

$$4NH_3(g) + 5O_2(g) \rightarrow 4NO(g) + 6H_2O(g)$$

The nitrogen monoxide reacts with oxygen in the air forming nitrogen dioxide.

nitrogen monoxide + oxygen $\rightarrow$ nitrogen dioxide

$$2NO(g) + O_2(g) \rightarrow 2NO_2(g)$$

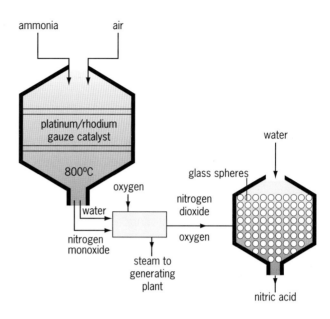

Manufacture of nitric acid

The nitrogen dioxide then dissolves in water forming nitric acid.

nitrogen dioxide + water $\rightarrow$ nitric acid + nitrogen monoxide

$$3NO_2(g) + H_2O(g) \rightarrow 2HNO_3(aq) + NO(g)$$

Ammonium nitrate – the fertiliser

The ammonia produced in the Haber process is then reacted with the nitric acid to produce the fertiliser, ammonium nitrate.

ammonia + nitric acid $\rightarrow$ ammonium nitrate

$$NH_3(g) + HNO_3(aq) \rightarrow NH_4NO_3(aq)$$

The ammonium nitrate is turned into granules which can be scattered easily on the soil.

★ THINGS TO DO

1 Make a list of the raw materials needed for the industrial preparation of ammonium nitrate.

2 What will be the effect of (a) using high temperatures, and (b) using catalysts on the rate at which each reaction takes place?

3 When nitrates are put onto the soil, only about half is used by plants. The other half runs from the soil into rivers, ponds and lakes. Nitrates move slowly through the soil – at a rate of

about one metre per year. The groundwater in most parts of the country is about 25–30 metres below the surface.

Regulations set by the European Commission say that drinking water should contain no more than 50 mg of nitrates per litre. Many people receive drinking water which exceeds this limit.

Why is it unfair to blame modern farming methods for the problem? Who will be affected by nitrogenous fertilisers used today?

Economy and production

The Haber process was initially developed to supply ammonia for explosives during the First World War. Today most is used in the manufacture of nitric acid and fertilisers. As with all industrial processes, steps must be taken to ensure that the ammonia is produced as economically as possible. Economic production methods help keep the cost of the products low.

The main costs of production are:

- buildings and their maintenance and machinery;
- wages for employees;
- obtaining or producing the raw materials.

The raw materials for the Haber process are:

- air (from which nitrogen is obtained);
- natural gas (from which hydrogen is obtained);
- water (needed for the steam which is used to produce hydrogen from natural gas);
- iron and nickel for the catalysts;
- electricity.

To keep production costs low, as much ammonia as possible must be made from a given amount of the raw materials. At each stage of the production process, steps are taken to increase the yield, and to recycle unused reactants.

Ammonia production

The ammonia is produced as nitrogen and hydrogen react exothermically.

$$\text{nitrogen} + \text{hydrogen} \rightleftharpoons \text{ammonia}$$
$$N_2(g) + 3H_2(g) \rightleftharpoons 2NH_3(g)$$
$$\Delta H = -92\,\text{kJ mol}^{-1}$$

In Topic 4.13 you saw that the reaction between nitrogen and hydrogen is reversible. This means that when the ammonia is produced some of it decomposes into nitrogen and hydrogen (the original reactants). So whilst the reaction proceeds, some ammonia is formed, and also some hydrogen and nitrogen are re-formed and recycled.

If the conditions of temperature and pressure are kept constant, the reaction continues at this rate, producing a continuous yield of 15%. This situation is called a **chemical equilibrium**.

Because the processes continue in both directions continually the equilibrium is described as a **dynamic equilibrium**.

Ammonia production plant

The position of equilibrium affects the yield. If the equilibrium position is to the left, then the yield is reduced (less nitrogen and hydrogen react to produce ammonia). If the equilibrium position is to the right, the yield is increased as more of the hydrogen and nitrogen react. To make the process economic the equilibrium position must be as far to the right as possible, ensuring that the yield is high.

Changing the yield

At room temperatures the rate of reaction between nitrogen and hydrogen is so slow that the process is uneconomic. As with other reactions, however, the rate of reaction increases with temperature. At higher temperatures, nitrogen and hydrogen react much faster, but this introduces another problem – the yield of ammonia decreases (because the equilibrium position is moved to the left).

In 1888 Henri Le Chatelier, a French scientist, suggested some general principles which affect the equilibrium position in reactions involving gases.

He found that if the pressure was increased, the reaction tended to take place faster towards the side which contained the fewest molecules of gas. In the Haber process, for example, there are four molecules on the left-hand side of the equation for the reaction but only two on the right.

Carrying out the reaction at higher pressure therefore moves the equilibrium position to the right, increasing the yield of ammonia.

He also noticed that exothermic reactions produced more of the product if the temperature was low. This poses another problem. If the process is carried out at room temperature a higher percentage of ammonia is produced but the rate of reaction is so slow that it makes the process very expensive – it becomes uneconomic.

We can, however, increase the rate of a reaction using a catalyst. The catalyst does not change the position of equilibrium in any way (the equilibrium position under any conditions would be the same with or without a catalyst), but speeds up the reaction so that more of the product is produced in any given time.

A compromise must be reached between the two conditions of temperature and pressure which produces a sensible, and economic yield of ammonia.

The graph shows how the yield of ammonia varies at different temperatures and pressures. You can see that:

- as the pressure increases at any given temperature, the yield of ammonia increases;
- as the temperature increases at any given pressure the yield of ammonia decreases.

In the case of ammonia production a compromise is reached where the reaction takes place inside a

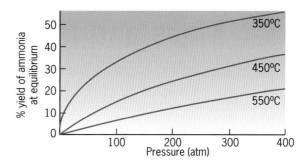

The yield of ammonia varies at different temperatures

steel vessel at a pressure of 200 atmospheres and at a temperature of between 350°C and 500°C – conditions which are, to some extent, limited by the economics of producing such high pressures and manufacturing vessels which can withstand them.

Inside the pressure vessel the gases pass over a catalyst of freshly produced, finely divided iron. (See diagram of Haber process in previous Topic.)

The gas mixture leaving the reaction vessel contains only about 15% ammonia which is removed by cooling and condensing. The remaining (unreacted) nitrogen and hydrogen are recycled and pass through the system again.

★ THINGS TO DO

1 Use the information given above and any other sources you may have to produce a flow diagram of the Haber process. Indicate the flow of the gases and write equation(s) to show what happens at each stage. For the stage where the ammonia is separated from the reaction vessel mixture you may have to look up the boiling points of the gases involved.

2 The table opposite shows some properties of the main gases in the air.

a) To what temperature would the air have to be lowered before it 'liquefied'? Would all the gases of the air be liquid at that temperature? If not, what would they be?

b) At what temperature would nitrogen be separated during the fractionation of air? Draw a fractionating column showing where each gas would be separated and the temperatures at which they would separate.

c) During manufacture, what special property of ammonia allows it to be separated from the nitrogen and hydrogen which are left over after passing through the converter?

d) In the manufacture of ammonia, when the nitrogen and hydrogen are heated under high pressures a catalyst is used. What effect will the catalyst have on what happens? What happens to the catalyst during the reaction?

Gas present	Percentage	Melting point (°C)	Boiling point (°C)
Argon	0.93%	−189	−186
Carbon dioxide	0.03%	sublimes	−78
Neon	0.002%	−248	−246
Nitrogen	78.09%	−210	−196
Oxygen	20.95%	−218	−183

Mass and moles

Chemists often need to know how much of a substance has been formed or used up during a chemical reaction. This is particularly important in the chemical industry where the substances you are reacting (the reactants) and the substances being produced (the products) are worth thousands of pounds. Waste costs money!

The control room at ICI Billingham – ICI chemists need to know how much fertiliser is going to be produced

A mole of aluminium

A mole of iron

To solve this problem we need a way of counting atoms, ions and molecules. Atoms, ions and molecules are very tiny particles and it is impossible to measure out a dozen or even one hundred of them. Instead chemists weigh out a very large number called a **mole** (often abbreviated to **mol**). A mole is 6×10^{23} atoms, ions or molecules. This number (6×10^{23}) is called **Avogadro's constant**, after a famous Italian scientist called Amadeo Avogadro who lived from 1776 to 1856.

So a mole of the element aluminium is 6×10^{23} atoms of aluminium and a mole of the element iron is 6×10^{23} atoms of iron.

Calculating moles

The average mass of a large number of atoms of an element is called its relative atomic mass (symbol A_r). This quantity takes into account the percentage abundance of all the isotopes of an element that exist.

In 1961 the International Union of Pure and Applied Chemistry (IUPAC) recommended that the standard used for the A_r scale should be carbon–12. An atom of carbon–12 was taken to have a mass of 12 amu. The A_r of an element is now defined as the average mass of its isotopes compared to one-twelfth the mass of one atom of carbon–12.

$$A_r = \frac{\text{average mass of isotopes of the element}}{\frac{1}{12} \times \text{mass of 1 atom of carbon–12}}$$

Note: $\frac{1}{12}$ of the mass of one carbon–12 atom = 1 amu.

For example, consider the case of chlorine, which has two isotopes.

	$^{35}_{17}\text{Cl}$	$^{37}_{17}\text{Cl}$
% abundance	75	25
Ratio of atoms	3	1

Hence the 'average mass' of a chlorine atom is

$$\frac{(3 \times 35) + (1 \times 37)}{4} = 35.5$$

$$A_r = \frac{35.5}{1}$$

$$= 35.5 \text{ amu}$$

Chemists have found by experiment that if you take the **relative atomic mass** (A_r) of an element in grams that it always contains 6×10^{23} atoms or one mole of its atoms.

Moles and elements

For example, the relative atomic mass (A_r) of iron is 56. In 56 g of iron it is found that there are 6×10^{23} atoms. Therefore 56 g of iron is a mole of iron atoms.

A_r for aluminium is 27. In 27 g of aluminium it is found that there are 6×10^{23} atoms. Therefore 27 g of aluminium is one mole of aluminium atoms.

The mass of a substance present in any number of moles can be calculated using the relationship:

$$\frac{\text{mass}}{\text{(in grams)}} = \text{number of moles} \times \frac{\text{mass of 1 mole}}{\text{of the element}}$$

Example
Calculate the mass of 2 moles of iron (A_r: Fe = 56).

$$\frac{\text{mass of}}{\text{2 moles of iron}} = \frac{\text{number}}{\text{of moles}} \times \frac{\text{relative atomic}}{\text{mass } (A_r)}$$

$$= 2 \times 56$$

$$= 112 \text{ g}$$

If we know the mass of the element then it is possible to calculate the number of moles of that element using:

$$\text{number of moles} = \frac{\text{mass of the element}}{\text{mass of 1 mole of that element}}$$

Example

Calculate the number of moles of aluminium present in 108 g of the element (A_r: Al = 27).

$$\text{number of moles of aluminium} = \frac{\text{mass of aluminium}}{\text{mass of 1 mole of aluminium}}$$

$$= \frac{108}{27}$$

$$= 4$$

Moles and compounds

The idea of the mole has been used so far only with elements and atoms. It can also be used with compounds.

We cannot discuss the atomic mass of a molecule or of a compound because more than one type of atom is involved. Instead we have to discuss the **relative formula mass** (RFM). This is the sum of the relative atomic masses of all those elements shown in the formula of the substance.

For example, what is the mass of one mole of water (H_2O) molecules? (A_r: H = 1; O = 16)

From the formula of water, H_2O, you will see that one mole of water molecules contains 2 moles of hydrogen (H) atoms and 1 mole of oxygen (O) atoms. The mass of one mole of water molecules is therefore:

$$(2 \times 1) + (1 \times 16) = 18 \text{g}$$

The mass of one mole of a compound is called its **molar mass**. If you write the molar mass of a compound without any units then it is the relative formula mass, often called the relative molecular mass (M_r). So the relative formula mass of water is 18.

Example

Ethanol has the formula C_2H_5OH. What is:

a) the mass of one mole?
b) the relative formula mass of ethanol?
(A_r: C = 12; H = 1; O = 16.)

a) One mole of C_2H_5OH contains: 2 moles of carbon atoms, 6 moles of hydrogen atoms and 1 mole of oxygen atoms. Therefore:

$$\text{mass of one mole of ethanol} = (2 \times 12) + (6 \times 1) + (1 \times 16)$$

$$= 46 \text{g}$$

b) The relative formula mass (RFM) of ethanol is 46.

The mass of a compound found in any number of moles can be calculated using the relationship:

$$\text{mass of compound} = \text{number of moles of the compound} \times \text{mass of 1 mole of the compound}$$

Example

Calculate the mass of 3 moles of carbon dioxide gas, CO_2. (A_r: C = 12; O = 16.)

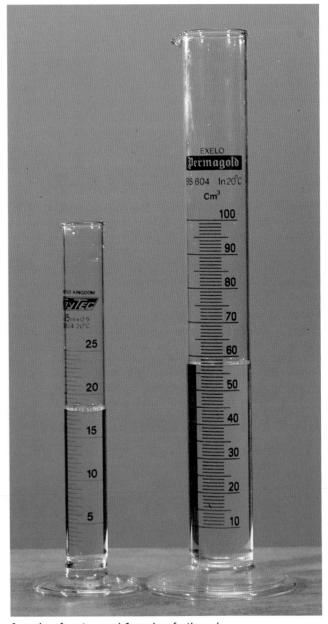

1 mole of water and 1 mole of ethanol

One mole of CO_2 contains: 1 mole of carbon atoms, 2 moles of oxygen atoms. Therefore:

$$\text{mass of 1 mole of } CO_2 = (1 \times 12) + (2 \times 16)$$
$$= 44\,g$$

$$\frac{\text{mass of 3}}{\text{moles of } CO_2} = \frac{\text{number}}{\text{of moles}} \times \frac{\text{mass of 1}}{\text{mole of } CO_2}$$

$$= 3 \times 44 = 132\,g$$

If we know the mass of the compound then it is possible to calculate the number of moles of the compound using the relationship:

$$\frac{\text{number of moles}}{\text{of compound}} = \frac{\text{mass of compound}}{\text{mass of 1 mole of compound}}$$

Example

Calculate the number of moles of magnesium oxide, MgO, in 80 g of the compound. (A_r: Mg = 24; O = 16.)

One mole of MgO contains: 1 mole of magnesium atoms, 1 mole of oxygen atoms. Therefore:

$$\text{mass of one mole of MgO} = (1 \times 24) + (1 \times 16)$$
$$= 40\,g$$

$$\frac{\text{number of moles}}{\text{of MgO in 80 g}} = \frac{\text{mass of MgO}}{\text{mass of 1 mole of MgO}}$$

$$= \frac{80}{40}$$

$$= 2$$

★ THINGS TO DO

Use the values of A_r which follow to answer the questions below: C = 12; Ne = 20; Mg = 24; O = 16; S = 32; Na = 23; H = 1; Fe = 56; Cu = 63.5; N = 14; Zn = 65; K = 39.

1 Calculate the number of moles in:
   a) 4 g of neon atoms;
   b) 8 g of magnesium atoms;
   c) 36 g of carbon atoms.

2 Calculate the mass of:
   a) 0.1 moles of nitrogen molecules;
   b) 6 moles of sulphur atoms;
   c) 0.25 moles of potassium atoms.

3 Calculate the number of moles in:
   a) 0.98 g of sulphuric acid (H_2SO_4);
   b) 400 g of sodium hydroxide (NaOH);
   c) 72 g of iron(II) oxide (FeO).

4 Calculate the mass of:
   a) 0.2 moles of zinc oxide (ZnO);
   b) 2.5 moles of hydrogen sulphide (H_2S);
   c) 0.45 moles of copper(II) sulphate ($CuSO_4$).

Moles, volumes and concentration

Moles and volumes

Many substances exist as gases. If we want to find the number of moles of a gas we can do this by measuring the mass but it is more usual to measure the volume.

Chemists have shown by experiment that one mole of any gas occupies a volume of approximately 24dm³ (24l) at room temperature and pressure (rtp).

This container has 230 g of butane in it. How many moles of butane is this?

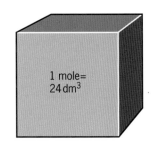

1 mole=
24 dm³

One mole takes up 24 dm³ at room temperature and pressure

Therefore, it is relatively easy to convert volumes of gases into moles and moles of gases into volumes using the following relationship:

$$\text{number of moles of a gas} = \frac{\text{volume of the gas (in dm}^3 \text{ at rtp)}}{24 \text{ dm}^3}$$

or

$$\text{volume of a gas (in dm}^3 \text{ at rtp)} = \text{number of moles of gas} \times 24 \text{ dm}^3$$

Examples

1 Calculate the moles of ammonia gas, NH_3, in a volume of 72 dm³ of the gas measured at rtp.

$$\text{number of moles of ammonia} = \frac{\text{volume of ammonia in dm}^3}{24 \text{ dm}^3}$$

$$= \frac{72}{24}$$

$$= 3$$

2 Calculate the volume of carbon dioxide gas, CO_2, occupied by 0.5 moles of the gas measured at rtp.

$$\text{volume of } CO_2 = \text{number of moles of } CO_2 \times 24 \text{ dm}^3$$

$$= 0.5 \times 24$$

$$= 12 \text{ dm}^3$$

The volume occupied by one mole of any gas must contain 6×10^{23} molecules. Therefore, it follows that equal volumes of all gases measured at the same temperature and pressure must contain the same number of molecules. This idea was first put forward by Avogadro in 1811 and is called **Avogadro's Law**.

Moles and solutions

Chemists often need to know the concentration of a solution. Concentration is measured in moles per dm³ ($mol \, dm^{-3}$). When one mole of a substance is dissolved in water and the solution made up to 1 dm³ (1000 cm³) a 1 molar (1 M) solution is produced. Chemists do not always need to make up such large volumes of solution. A simple method of calculating the concentration is by using the relationship:

$$\text{concentration} = \frac{\text{number of moles}}{\text{volume (in dm}^3)}$$

Example
Calculate the concentration (in $mol \, dm^{-3}$) of a solution of sodium hydroxide, NaOH, which was made by dissolving 10 g of solid sodium hydroxide in water and making up to 250 cm³. (A_r: Na = 23; O = 16; H = 1.)

1 mole of NaOH contains: 1 mole of sodium, 1 mole of oxygen, 1 mole of hydrogen. Therefore

$$\text{mass of one mole of NaOH} = (1 \times 23) + (1 \times 16) + (1 \times 1)$$

$$= 40 \text{ g}$$

$$\text{number of moles} \atop \text{of NaOH in 10 g} = \frac{\text{mass of NaOH}}{\text{mass of 1 mole of NaOH}}$$

$$= \frac{10}{40}$$

$$= 0.25$$

$$[250\,\text{cm}^3 = \frac{250}{1000}\,\text{dm}^3 = 0.25\,\text{dm}^3]$$

$$\text{concentration of} \atop \text{NaOH solution} = \frac{\text{number of moles of NaOH}}{\text{volume of solution (in dm}^3)}$$

$$= \frac{0.25}{0.25}$$

$$= 1\,\text{mol dm}^3 \text{ (or 1 M)}$$

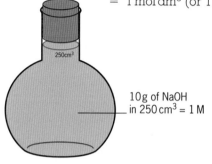

10g of NaOH
in 250 cm³ = 1 M

Volumetric flasks are used to make solutions of accurate concentration

Sometimes chemists need to know the mass of a substance that has to be dissolved to prepare a known volume of solution at a given concentration.

A simple method of calculating the number of moles and hence the mass of substance needed is by using the relationship:

$$\text{number} \atop \text{of moles} = {\text{concentration} \atop \text{(in mol dm}^{-3})} \times {\text{volume of solution} \atop \text{(in dm}^3)}$$

Example

Calculate the mass of potassium hydroxide, KOH, that needs to be used to prepare 500 cm³ of a 2 mol dm⁻³ (2 M) solution in water. (A_r: K = 39; O = 16; H = 1.)

$$\text{number of} \atop \text{moles of KOH} = {\text{concentration of} \atop \text{solution (mol dm}^{-3})} \times {\text{volume of} \atop \text{solution (dm}^3)}$$

$$= 2 \times \frac{500}{1000}$$

$$= 1$$

1 mole of KOH contains: 1 mole of potassium, 1 mole of oxygen and 1 mole of hydrogen. Therefore,

$$\text{mass of 1} \atop \text{mole of KOH} = (1 \times 39) + (1 \times 16) + (1 \times 1)$$

$$= 56\,g$$

Therefore

$$\text{mass of KOH} \atop \text{in one mole} = {\text{number} \atop \text{of moles}} \times {\text{mass of} \atop \text{1 mole}}$$

$$= 1 \times 56$$

$$= 56\,g$$

★ THINGS TO DO

Use the values of A_r which follow to answer the questions below: C = 12; Ne = 20; Mg = 24; O = 16; S = 32; Na = 23; H = 1; Fe = 56; Cu = 63.5; N = 14; Zn = 65; K = 39
   1 mole of any gas at room temperature and pressure occupies 24 dm³.

1 Calculate the number of moles at room temperature and pressure in:
   a) 4 dm³ of carbon dioxide (CO_2);
   b) 24 dm³ of sulphur dioxide (SO_2);
   c) 12 cm³ of carbon monoxide (CO).

2 Calculate the volume of:
   a) 0.2 moles of hydrogen chloride (HCl);
   b) 2.2 g of carbon dioxide (CO_2);
   c) 340 g of ammonia (NH_3).

3 Calculate the concentration of solutions containing:
   a) 0.3 moles of sodium hydroxide dissolved in water and made up to 100 cm³;
   b) 4.9 g of sulphuric acid (H_2SO_4) dissolved in water and made up to 500 cm³.

4 Calculate the mass of:
   a) copper(II) sulphate ($CuSO_4$) which needs to be used to prepare 500 cm³ of a 0.2 M solution;
   b) potassium nitrate (KNO_3) which needs to be used to prepare 200 cm³ of a 4 M solution.

Calculating formulae

If we have one mole of a compound then the formula shows the number of moles of each element in that compound. For example, the formula for lead(II) bromide is $PbBr_2$. This means that 1 mole of lead(II) bromide contains 1 mole of lead and 2 moles of bromine. If we do not know the formula of a compound we can find the masses of the elements present experimentally and these masses can be used to work out the formula of that compound.

Finding the formula of magnesium oxide

When magnesium ribbon is heated strongly it burns very brightly to form the white powder called magnesium oxide.

$$magnesium + oxygen \rightarrow magnesium\ oxide$$
$$2Mg(s) + O_2(g) \rightarrow 2MgO(s)$$

The following data was obtained from an experiment using the apparatus shown in the illustration to obtain the formula for this white powder, magnesium oxide.
Experimental data:

Mass of crucible = 14.63 g
Mass of crucible and magnesium = 14.87 g
Mass of crucible and magnesium oxide = 15.03 g
Mass of magnesium used = 0.24 g
Mass of oxygen which has reacted with the magnesium = 0.16 g

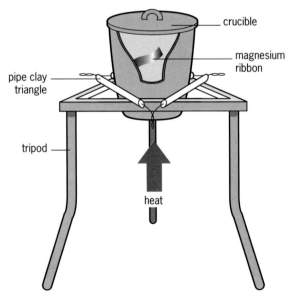

The apparatus used to find out the formula of magnesium oxide

From this data we can calculate the number of moles of each of the reacting elements. (A_r: Mg = 24; O = 16.)

	Mg	O
Masses reacting (g)	0.24	0.16
Number of moles	$\dfrac{0.24}{24}$	$\dfrac{0.16}{16}$
	= 0.01	= 0.01
Ratio of moles	1	1
Formula	MgO	

This formula is called the **empirical formula** of the compound. It shows the simplest ratio of the atoms present.

Finding the formula of an organic compound

In another experiment an unknown organic compound was found to contain 0.12 g of carbon and 0.02 g of hydrogen. Calculate the empirical formula of the compound. (A_r: C = 12; H = 1.)

	C	H
Masses (g)	0.12	0.02
Number of moles	$\dfrac{0.12}{12}$	$\dfrac{0.02}{1}$
	= 0.01	= 0.02
Ratio of moles	1	2
Empirical formula	CH_2	

However, from our knowledge of bonding (see Topic 3.12) a molecule of this formula cannot exist, but molecules of the following formulae do exist: C_2H_4, C_3H_6, C_4H_8, C_5H_{10}. All of these formulae show the same ratio of carbon atoms to hydrogen atoms, CH_2, as our example. To find out which of these formulae is the actual formula for the unknown organic compound we need to know the mass of one mole of the compound.

Using the mass spectrometer shown in the illustration, the relative molecular mass (M_r) of this organic compound was found to be 56. We need to find out the number of empirical formulae units present.

$$\text{M_r of the empirical formula unit} = (1 \times 12) + (2 \times 1) = 14\,g$$

$$\text{Number of empirical formula units present} = \frac{\text{M_r of the compound}}{\text{M_r of empirical formula unit}}$$

$$= \frac{56}{14}$$

$$= 4$$

Empirical formula (from experiment) = CH_2
M_r of organic compound = 56 M_r of CH_2 = 14
Molecular formula is $(CH_2)_4 = C_4H_8$

Therefore, the actual formula of the unknown organic compound is $4 \times CH_2 = C_4H_8$.

This substance is called butene. C_4H_8 is the molecular formula for this substance and shows the actual numbers of atoms of each element present in one molecule of the substance.

Sometimes the composition of a compound is given as a percentage by mass of the elements present. In cases such as this the procedure shown in the following example is followed.

Example
a) Calculate the empirical formula of an organic compound which contains 92.3% carbon and 7.7% hydrogen by mass.
b) The M_r of the organic compound is 78. What is its molecular formula? (A_r: C = 12; H = 1.)

A mass spectrometer measures the relative masses of the substances placed in it

a)

	C	H
% by mass	92.3	7.7
in 100 g	92.3 g	7.7 g
moles	$\dfrac{92.3}{12}$	$\dfrac{7.7}{1}$
	= 7.7	= 7.7
Ratio of moles	1	1
Empirical formula	CH	

b) M_r of the empirical formula unit
CH = 12 + 1 = 13

$$\text{Number of empirical formula units present} = \frac{\text{M_r of compound}}{\text{M_r of empirical formula unit}}$$

$$= \frac{78}{13}$$

$$= 6$$

The molecular formula of the organic compound is $6 \times CH = C_6H_6$.

Empirical formula (from experiment) = CH
M_r of organic compound = 78 M_r of CH = 13
Molecular formula is $(CH)_6 = C_6H_6$

★ THINGS TO DO

Use the following values of A_r to answer the questions below: H = 1; C = 12; O = 16; Ca = 40.

1 Determine the empirical formula of an oxide of calcium formed when 0.4 g of calcium reacts with 0.16 g of oxygen.

2 Determine the empirical formula of an organic hydrocarbon compound which contains 80% by mass of carbon and 20% by mass of hydrogen. If the M_r of the compound is 30 what is its molecular formula?

Moles and chemical equations

When we write a balanced chemical equation we are indicating the numbers of moles of reactants and products involved in the chemical reaction. Consider the reaction between magnesium and oxygen.

magnesium + oxygen → magnesium oxide
$$2Mg(s) + O_2(g) → 2MgO(s)$$

This shows that 2 moles of magnesium react with 1 mole of oxygen to give 2 moles of magnesium oxide.

Using the ideas of moles and masses we can use this information to calculate the quantities of the different chemicals involved.

$2Mg(s)$ +	$O_2(g)$ →	$2MgO(s)$
2 moles	1 mole	2 moles
2×24	$1 \times 16 \times 2$	$2 \times (24 + 16)$
$= 48\,g$	$= 32\,g$	$= 80\,g$

You will notice that the total mass of reactants is equal to the total mass of product. This is true for any chemical reaction and it is known as the **law of conservation of mass** for a chemical reaction.

Chemists use this idea to calculate masses of products formed and reactants used in chemical processes before they are carried out.

Example using a solid

Lime (calcium oxide, CaO) is used in the manufacture of mortar. It is manufactured in large quantities by heating limestone (calcium carbonate, $CaCO_3$). The equation for the process is:

$CaCO_3(s)$	→ $CaO(s)$ +	$CO_2(g)$
1 mole	1 mole	1 mole
$(40 + 12 + (3 \times 16))$	$40 + 16$	$12 + (2 \times 16)$
$= 100\,g$	$= 56\,g$	$= 44\,g$

Calculate the amount of lime produced when 10 tonnes of limestone are heated. (A_r: Ca = 40; C = 12; O = 16.)

1 tonne (t) = 1000 kg   1 kg = 1000 g

From this relationship between grams and tonnes we can replace the masses in grams by masses in tonnes.

$CaCO_3(s)$ →	$CaO(s)$ +	$CO_2(g)$
100 t	56 t	44 t
10 t	5.6 t	4.4 t

The lime kiln operator needs to know the number of moles of $CaCO_3$ to add to just remove all the impurities

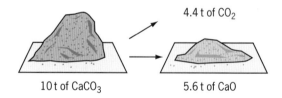

4.4 t of CO_2

10 t of $CaCO_3$ 5.6 t of CaO

The equation now shows that 100 t of limestone will produce 56 t of lime. Therefore, 10 t of limestone will produce 5.6 t of lime.

Example using a gas

Many chemical processes involve gases. The volume of a gas is measured more easily than its mass. The next example shows how chemists work out the volumes of gaseous reactants and products needed using ideas of moles.

Some rockets use hydrogen gas as a fuel. When hydrogen burns in oxygen it forms steam. Calculate the volumes of **a)** water ($H_2O(g)$) produced and **b)** $O_2(g)$ used if 960 dm³ of hydrogen gas ($H_2(g)$) were burned in oxygen. (A_r: H = 1; O = 16.) Assume 1 mole of any gas occupies a volume of 24 dm³.

$$2H_{2(g)} \quad + \quad O_{2(g)} \quad \rightarrow \quad 2H_2O_{(g)}$$

2 moles	1 mole	2 moles
2×24	1×24	2×24
$= 48\,dm^3$	$= 24\,dm^3$	$= 48\,dm^3$

Therefore

	$96\,dm^3$	$48\,dm^3$	$96\,dm^3$ ($\times 2$)
so	$960\,dm^3$	$480\,dm^3$	$960\,dm^3$ ($\times 10$)

When 960 dm^3 of hydrogen are burned in oxygen:

a) 480 dm^3 of oxygen are required and

b) 960 dm^3 of H$_2$O(g) are produced.

Examples using a solution

Chemists usually carry out chemical reactions using solutions. If they know the concentration of the solution(s) they are using then they can find out the quantities reacting.

Examples

1 Calculate the volume of 1 M H$_2$SO$_4$ required to react completely with 6 g of magnesium. (A_r: Mg = 24.)

$$\frac{\text{number of moles}}{\text{of magnesium}} = \frac{\text{mass of magnesium}}{\text{mass of 1 mole of magnesium}}$$

$$= \frac{6}{24} = 0.25$$

$$Mg_{(s)} + H_2SO_{4(aq)} \rightarrow MgSO_{4(aq)} + H_{2(g)}$$

1 mole	1 mole	1 mole	1 mole
0.25 mol	0.25 mol	0.25 mol	0.25 mol

We can see that 0.25 mol of H$_2$SO$_4$(aq) are required. Using:

$$\frac{\text{volume of}}{\text{H}_2\text{SO}_{4}\text{(aq) (dm}^3)} = \frac{\text{moles of H}_2\text{SO}_4}{\text{conc of H}_2\text{SO}_4 \text{ (mol dm}^{-3})}$$

$$= \frac{0.25}{1} = 0.25\,dm^3 \text{ or } 250\,cm^3$$

2 What is the concentration of sodium hydroxide solution used in the following neutralisation reaction? 40 cm^3 of 0.2 M hydrochloric acid just neutralised 20 cm^3 of sodium hydroxide solution.

$$\frac{\text{Number of moles}}{\text{of HCl used}} = \text{conc (mol dm}^{-3}) \times \text{volume (dm}^3)$$

$$= 0.2 \times 0.004 = 0.008$$

$$HCl_{(aq)} + NaOH_{(aq)} \rightarrow NaCl_{(aq)} + H_2O_{(l)}$$

1 mole	1 mole	1 mole	1 mole
0.008 mol	0.008 mol	0.008 mol	0.008 mol

You will see that 0.008 moles of NaOH would have been present. The concentration of the NaOH(aq) will be given by:

$$\frac{\text{concentration of}}{\text{NaOH (mol dm}^{-3})} = \frac{\text{number of moles of NaOH}}{\text{volume of NaOH (dm}^3)}$$

$$\left(\frac{\text{volume of}}{\text{NaOH in dm}^3} = \frac{20}{1000} = 0.02 \right)$$

$$= \frac{0.008}{0.02} = 0.4\,\text{mol dm}^{-3} \text{ or } 0.4M$$

★ THINGS TO DO

Use the following A_r values to answer the questions below: O = 16; S = 32; Cu = 63.5; O = 16; Mg = 24; K = 39.

1 Calculate the mass of sulphur dioxide produced by burning 16 g of sulphur in an excess of oxygen in the Contact process.

2 Calculate the mass of sulphur which when burned in excess oxygen produces 640 g of sulphur dioxide in the Contact process.

3 Calculate the mass of copper required to produce 159 g of copper(II) oxide when heated in excess oxygen.

4 In the rocket mentioned in the example using a gas, in which hydrogen is used as a fuel, calculate the volume of hydrogen used to produce 24 dm^3 of water (H$_2$O(g)).

5 Calculate the volume of 2 M sulphuric acid required to react with 8 g magnesium.

6 What is the concentration of potassium hydroxide solution used in the following neutralisation reaction? 20 cm^3 of 0.2 M hydrochloric acid just neutralised 15 cm^3 of potassium hydroxide solution.

Exam questions

denotes higher level questions

1 Some students have studied the reaction between magnesium and dilute hydrochloric acid.
   a) They carried out an investigation to discover whether the concentration of acid used had any effect on the time taken for the magnesium to react completely. Equal volumes of acid and equal lengths of magnesium ribbon were used in each test.
   The table shows a set of results.

Concentration of acid/moles per dm³	Time taken for magnesium to react completely /seconds
2.0	13
1.5	22
1.2	30
0.8	70
0.6	145
0.5	250

i) Plot the points on the grid and draw a line graph to show the changes in the time taken for the reaction to finish, when different concentrations of acid are used. (2)

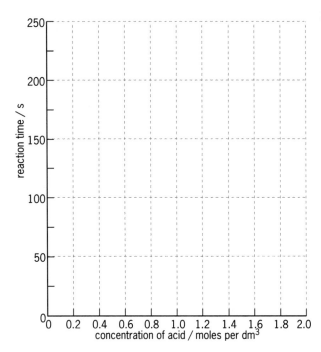

ii) Use your graph to help you to describe how increasing the concentration of acid affects the rate of reaction. (1)
iii) Use your graph to predict the time taken for the reaction if the concentration of acid was 1.0 moles per dm³. (1)
iv) Write down **two** other factors which could change the rate of this reaction. In each case state how the rate would change. (2)
v) Use the idea of particles to explain the effect of changing concentration on the rate of reaction. (2)
b) One student asked if the class could repeat their experiment using potassium instead of magnesium. The teacher said 'Definitely not!'
   Use your knowledge of the reactivity series to explain why the teacher was very wise to say this. (2)

c) The equation for the reaction between magnesium and hydrochloric acid is

$$Mg + 2HCl \rightarrow MgCl_2 + H_2$$

Calculate the volume of hydrogen produced when 0.5 g of magnesium reacts with an excess of hydrochloric acid. (Assume all measurements are made at room temperature and atmospheric pressure.) (Answer in dm³ hydrogen gas.) (4)
(MEG, 1995)

2 Yeast cells can be used to change sugar to alcohol. During this process the yeast grows. This process can be shown by the word equation:

Sugar $\xrightarrow{\text{Yeast}}$ Energy + Alcohol + Carbon dioxide

a) i) Which substance produced in this process is important in wine-making? (1)
ii) Which substance produced in this process is important in bread-making? (1)
b) A sweet factory produces waste water containing sugar. The water will go into a local river.
   A local scientist says they should not do this before treating the waste sugar. She suggests that they should put yeast in it. She tells them that this could be an advantage and that yeast can be sold as an animal food.
   Why does the scientist tell them not to put the waste water in the river? Explain how her advice would help the factory to deal with the problem. (4)
(ULEAC, 1995)

3 The symbol equation below shows the reaction when methane burns in oxygen.

$$CH_4 + 2O_2 \rightarrow CO_2 + 2H_2O$$

a) How much oxygen is needed to burn $100\,cm^3$ of methane? (Answer in cm^3) (1)

b) An energy level diagram for this reaction is shown below.

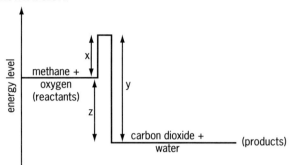

i) Which chemical bonds are broken and which are formed during this reaction? (4)

ii) Explain the significance of x, y and z on the energy level diagram in terms of the energy transfers which occur when these chemical bonds are broken and formed. (5)

(NEAB, Specimen)

4 To obtain ammonia from nitrogen and hydrogen in the reaction:

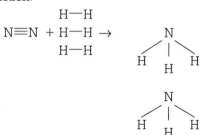

bonds in nitrogen and hydrogen need to be broken and bonds in ammonia have to be formed. Use **[page 16 of]** the **Data Book** to calculate the relative energy needed to:

a) i) break **all** the bonds in the reactants;

ii) form **all** the bonds in the products. (3)

b) Use your results from part **(a)** to explain why the reaction is exothermic. (2)

c) Indicate on the diagram below, the energy level of the ammonia relative to that of the reactants. (1)

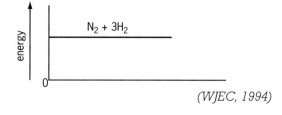

(WJEC, 1994)

5 In a reaction known as 'slaking' water is added to calcium oxide to produce calcium hydroxide. The calcium hydroxide is then used to make plaster.

$$CaO + H_2O \rightarrow Ca(OH)_2$$

The diagram represents the energy change during this reaction.

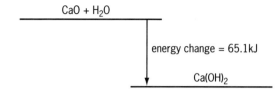

a) What does the diagram tell us about the energy change which takes place in this reaction? (2)

b) What does the diagram indicate about the relative amount of energy required to break bonds and form the new bonds in this reaction? (3)

(NEAB, Specimen)

6 Ammonia is manufactured in the Haber Process, from nitrogen and hydrogen.

a) Balance this symbol equation for the process.

$$N_2 + H_2 \rightleftharpoons NH_3 \qquad (2)$$

b) The graph below shows the percentage of reacting gases converted into ammonia, at different temperatures and pressures.

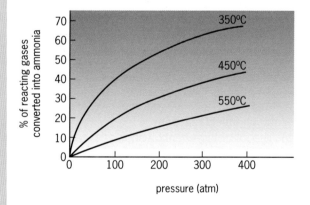

i) What does the graph suggest about the temperature and pressure needed to convert the maximum percentage of reacting gases into ammonia? (2)

ii) Suggest reasons why the manufacture of ammonia in the Haber process is usually carried out at about 400 °C and 200 atmospheres pressure. (2)

(NEAB, 1994)

7 **a)** Nitric acid, HNO_3 is made from ammonia by the Ostwald Process. There are three stages in this Process.

Stage 1: ammonia reacts with oxygen, O_2, from the air in the presence of a catalyst to form nitrogen oxide, NO.

Stage 2: nitrogen oxide and oxygen react to form nitrogen dioxide, NO_2.

Stage 3: nitrogen dioxide, oxygen and water react together to form nitric acid.

The equation for stage 1 is:

$$4NH_3 + 5O_2 \rightarrow 4NO + 6H_2O$$

i) Name the catalyst used in stage 1. (1)

ii) Complete and balance the equations for the reactions taking place in stages 2 and 3.

Stage 2: $2NO + O_2 \rightarrow \ldots$ (1)

Stage 3: $\ldots NO_2 + \ldots H_2O + O_2 \rightarrow \ldots HNO_3$ (1)

b) Ammonium nitrate is an important fertiliser. It is made by reacting nitric acid with the alkali ammonia.

i) State the type of reaction taking place. (1)

ii) The equation for this reaction is:

$$NH_3 + HNO_3 \rightarrow NH_4NO_3$$

Calculate the number of tonnes of ammonium nitrate that can be made from 68 tonnes of ammonia. (Relative atomic masses: H = 1, N = 14, O = 16) (3)

(NEAB, 1995)

8 a) An investigation was carried out to see how the rate of decomposition of hydrogen peroxide could be changed. The reaction is,

hydrogen peroxide $\rightarrow$ oxygen + water

Manganese(IV) oxide was added to the hydrogen peroxide.

What effect does the manganese(IV) oxide have on the chemical reaction? (1)

b) The following equation shows the reaction taking place between dilute hydrochloric acid and marble chips (calcium carbonate):

$$CaCO_{3(s)} + 2HCl_{(aq)} \rightarrow CaCl_{2(aq)} + H_2O_{(l)} + CO_{2(g)}$$

i) Give **two** ways of increasing the rate of this reaction **other** than by using a catalyst or by heating. (2)

ii) Calculate the mass of carbon dioxide produced if 25 g of calcium carbonate is reacted with excess acid. (Relative Atomic Masses: Calcium 40, Carbon 12, Oxygen 16.) (3)

c) In the Haber Process, ammonia (NH_3) is produced by reacting nitrogen (N_2) and hydrogen (H_2).

i) Write a balanced symbol equation for this reaction. (1)

ii) The reaction is reversible. What would happen to the equilibrium concentration of ammonia if the pressure of the equilibrium mixture was increased. Explain your answer. (2)

d) The reaction of methane and steam at 700 °C produces carbon monoxide and hydrogen,

$$CH_{4(g)} + H_2O_{(g)} \rightleftharpoons CO_{(g)} + 3H_{2(g)}$$
$$\Delta H = -207\,kJ \text{ per mole of methane}$$

i) What mass of methane would be needed in order to release 207 kJ of energy? (Relative Atomic Masses: Carbon 12, Hydrogen 1, Oxygen 16.) (2)

ii) Predict what will happen to the equilibrium concentration of hydrogen if the temperature is raised. Explain your answer. (2)

(SEG, 1994)

9 a) Analysis of a sample of the compound sodium oxide, showed that it contained 2.3 g of sodium and 0.8 g of oxygen.

Use this information to work out the formula of the compound sodium oxide.

(You may find it helpful to use the table below to set out your answer but use any other method if you prefer.)

	SODIUM	OXYGEN
Mass of each element in the sodium oxide		
Mass of 1 mole of each element		
Number of moles of elements combining		
Simplest ratio		
Formula of sodium oxide is		

(4)

b) *You may find it helpful to use the [information on page 7 of the Data Book] in answering this question.*

Sodium reacts with water to produce sodium hydroxide and hydrogen.

$$2Na_{(s)} + 2H_2O_{(l)} \rightarrow 2NaOH_{(aq)} + H_{2(g)}$$

Calculate the volume of hydrogen that would be produced at ordinary temperature and pressure from 3.5 g of sodium. (Show your working.) (5)

(NEAB, 1995)

Glossary

Acid   Acids dissolve in water producing $H^+(aq)$ ions forming a solution with a pH of less than 7.

Acid rain   Formed from acid gases (e.g. sulphur dioxide and nitrogen oxides) in the air dissolving in falling rain to produce an acidic solution.

Activation energy   The energy that particles of substances must have before they can take part in a particular chemical reaction.

Addition polymerisation   The formation of a polymer by an addition reaction e.g. polyethene is formed from the monomer ethene.

Aerobic respiration   The process by which living organisms take in oxygen from the atmosphere to oxidise their food to obtain energy.

Alcohols   These are organic compounds containing the –OH group. They have the general formula $C_nH_{2n+1}OH$. Ethanol is by far the most important of the alcohols and is usually just called 'alcohol'.

Algae   Simple plants which do not have roots, stems or leaves. Usually green but sometimes brown or red. Live mainly in water.

Alkali   A soluble base which dissolves to produce $OH^-(aq)$ forming a solution of pH greater than 7.

Alkali metals   The very reactive metals found in group 1 of the periodic table.

Alkaline earth metals   The reactive metals found in group 2 of the periodic table.

Alkanes   A family of saturated hydrocarbons with the general formula, C_nH_{2n+2}. The term 'saturated', in this context, is used to describe molecules that have only single bonds.

Alkenes   A family of unsaturated hydrocarbons with a general formula, C_nH_{2n}. The term 'unsaturated', in this context, is used to describe molecules which contain one or more double carbon–carbon bonds.

Allotropes   Different structural forms of the same element both having the same physical state. For example, carbon exists as the allotropes diamond, graphite (and buckminsterfullerene).

Alloy   A mixture of two or more metals. It is formed by mixing molten metals thoroughly.

Anaerobic respiration   The process by which organisms obtain energy from chemically combined oxygen when they do not have access to free oxygen.

Anode   The positive electrode.

Artificial fertiliser   A substance which is added to soil to increase the amount of elements such as nitrogen, potassium and phosphorus. This enables crops grown in the soil to grow more healthily and to produce higher yields.

Atmosphere   A mixture of gases that cloak the Earth.

Atomic mass unit (amu)   1 amu = $\frac{1}{12}$ the mass of an atom of the most abundant isotope of carbon ($^{12}_6C$).

Atomic number (proton number)   The number of protons in the nucleus of an atom. The number of electrons present in an atom. The order of the element within the periodic table. Given the symbol Z.

Atoms   The smallest stable part of an element.

Avogadro's constant   The number of particles in a mole of any substance, equal to 6.02×10^{23}.

Avogadro's law   Equal volumes of all gases contain equal numbers of molecules under the same conditions of temperature and pressure.

Bacteria   Single cells each with a wall but no proper nucleus. Occur in air, water, soil or inside other organisms. Many of them cause disease.

Base   A substance which neutralises an acid producing a salt and water as the only products.

Biotechnology   This involves making use of micro-organisms in commercial processes. For example, the process of fermentation is brought about by the enzymes in yeast.

Boiling point   The temperature at which the vapour pressure of a liquid is equal to that of the atmosphere.

Bond   A force which holds groups of atoms or ions together.

Bond energy   The amount of energy required to break one mole of a given covalent bond.

Canyon   A gorge or ravine formed by the erosion caused by a river.

Catalyst   A substance which alters the rate of a chemical reaction, usually speeding it up, without itself being chemically changed.

Catalytic converter   A device for converting dangerous exhaust gases from cars into less dangerous emissions. For example, carbon monoxide gas is converted to carbon dioxide gas.

Cathode   The negative electrode.

Change in enthalpy (enthalpy change)   The change in energy in going from reactants to products; given the symbol ΔH.

Charge   Can be either positive or negative. Unlike charges attract whilst like charges repel.

Chemical change   A permanent change in which a new substance is formed.

Chemical equation   A way of showing the changes which take place during a chemical reaction.

Chemical equilibrium   It is dynamic. The concentrations of the reactants and products remain constant during a reaction because the rate at which the forward reaction occurs is the same as that of the back reaction.

Chlor-alkali industry   The industry based around the electrolysis of brine (saturated sodium chloride solution).

Chlorophyll   A green pigment found in chloroplasts which traps the light energy and converts it to chemical energy during photosynthesis.

Clock reactions   Those reactions which produce a rapid change of colour.

Combustion   A chemical reaction in which a substance reacts rapidly with oxygen with the production of heat and light.

Competition reactions   Reactions in which metals compete for oxygen or anions. The more reactive metal: (i) takes the oxygen from the oxide of a less reactive metal; (ii) displaces the less reactive metal from a solution of that metal salt (a displacement reaction).

Complete combustion   The complete oxidation of a hydrocarbon molecule to form carbon dioxide and water only.

Compound   A substance formed by the chemical bonding of two or more elements in fixed proportions.

Compressed   The squashing together of the particles within a substance caused by an increase in pressure.

Condense   To change from a vapour or a gas into a liquid.

Continental crust   The Earth's crust under the continents.

Contract   To become smaller.

Convection currents   The transference of heat through a liquid or gas by the actual movement of the substance.

Core   The central part of the Earth, under the mantle, consisting of mainly iron and nickel.

Corrosion   The name given to the process which takes place when metals and alloys are chemically attacked by oxygen, water or any other substances found in their immediate environment.

Covalent bond   A chemical bond formed by the sharing of one or more pairs of electrons between two non-metal atoms.

Cracking   The process of breaking large molecules into smaller, more useful molecules. For example, the decomposition of a higher alkane into alkenes and alkanes each of lower relative molecular mass. The process involves passing the larger alkane molecule over a catalyst

of aluminium and chromium oxides, heated to 500°C.

Cross-linked   Covalent bonds linking different polymer chains and therefore increasing the rigidity of the plastic. Thermosetting plastics are usually heavily cross-linked.

Crude oil   A fossil fuel formed from the partly decayed remains of marine animals. Consists largely of hydrocarbons.

Crust   The outer layer of the Earth.

Crystal lattice (ionic lattice)   A regular 3-D arrangement of atoms/ions in a crystalline solid.

Crystalline   Having the ordered characteristics of crystals.

Crystallize   When crystals form from a solution during the process of crystallization.

Delta   The area of deposited sediment at the mouth of a river.

Density   The mass per unit volume of a substance.

Deposition   The laying down of sediments.

Diatomic molecules   A molecule containing two atoms e.g. hydrogen (H_2), oxygen (O_2).

Diffusion   The process by which different substances mix as a result of the random motions of their particles.

Disinfect   To remove micro-organisms by the use of chemicals.

Displacement reaction   A reaction in which a more reactive element displaces a less reactive element from solution.

Dissolve   To go into solution.

Double covalent bond   A chemical bond formed by the sharing of two pairs of electrons between two non-metal atoms.

Ductile   The property of a metal which enables it to be drawn out into a wire.

Dynamic equilibrium   An equilibrium during a chemical reaction in which both the forward and back reactions occur at the same time.

Effervescence   Fizzing caused by bubbles of a gas being formed within a liquid during a chemical reaction.

Electrode   A point where the electric current enters and leaves the electrolytic cell.

Electrolysis   A process in which a chemical reaction is caused by the passage of an electric current.

Electrolyte   A substance which will carry electric current only when it is molten or dissolved.

Electron configuration (electron structure)   A shorthand method of describing the arrangement of electrons within the energy levels of an atom.

Electron shells (energy levels)   The allowed energies of electrons in atoms.

Electron   A fundamental sub-atomic particle, with a negative charge, present in all atoms within energy levels around the nucleus.

Electroplating   The covering of objects with a thin layer of a metal using the process of electrolysis.

Electrostatic forces of attraction   A strong force of attraction between unlike charges.

Element   A substance which cannot be further divided into simpler substances by chemical methods.

Empirical formula   This shows the simplest ratio ot atoms present in a substance.

Endothermic reaction   A chemical reaction which absorbs heat energy from its surroundings.

Enthalpy   Energy stored in chemical bonds, given the symbol H.

Enthalpy of combustion   The enthalpy change which takes place when one mole of a substance is completely burned in oxygen.

Enzyme   A protein molecule produced in living cells. It acts as a biological catalyst and is specific to a certain reaction. It only operates within narrow temperature and pH ranges.

Epoch   A point in geological time which begins a new era.

Era   A period of geological time considered as being of a distinctive character.

Erosion   The wearing away of rocks and other deposits on the surface of the Earth by the action of wind, water and ice.

Evaporation   A process occurring at the surface of a liquid and involving the change of state of a liquid into a vapour at a temperature below the boiling point.

Exothermic reaction   A chemical reaction in which heat energy is produced and released to its surroundings.

Expanding   Becoming larger.

Extrusive igneous rocks   Igneous rocks formed quickly when magma bursts through the Earth's surface.

Faults   They are caused by the movement along fractures in the Earth's crust.

Fermentation   A reaction in which sugar is changed into alcohol and carbon dioxide by the action of enzymes in yeast.

Fertiliser   A substance which is added to the soil to increase its productivity.

Fibres   A natural or synthetic filament which may be spun into a yarn.

Filtration   The process of separating a solid from a liquid. It is done by the use of a fine filter paper which does not allow the solid to pass through.

Fissures   Long, narrow cracks in rock.

Fold mountains   Mountains caused by layers of rock becoming folded by forces within the Earth's crust.

Fossil fuels   Fuels, such as coal, oil and natural gas, formed from the remains of plants and animals.

Fossils   These are remains often found in sedimentary rocks of organisms deposited before the rock was formed.

Fractional distillation   A method of separating a mixture of miscible liquids whose boiling points are quite close together.

Fractions   These are the simpler mixtures or single substances obtained from fractional distillation.

Friction   The name given to the forces providing resistance to the motion of two materials passing over each other.

Fuels   Substances which produce heat energy when they are burned.

Galvanising   The process of coating a metal with zinc.

Gas   The physical state in which the particles of a substance have so little attractive forces between them that no regular arrangement is seen.

Geological time   Scale used to describe the different periods in the passing of time.

Giant ionic structures   A lattice held together by the electrostatic forces of attraction between ions.

Giant molecular structure   A molecule containing thousands of atoms per molecule.

Glacier   A slowly moving mass of ice.

Greenhouse effect   The absorption of infra-red radiation by gases such as carbon dioxide (a greenhouse gas).

Groups   A vertical column of the periodic table containing elements with similar properties with the same number of electrons in their outer energy levels. The elements have an increasing number of inner energy levels down the group.

Haber process   The industrial manufacture of ammonia gas from nitrogen and hydrogen.

Halogens   The non-metallic elements found in group 7 of the periodic table.

Heat of reaction   The energy given out or taken in when a chemical reaction occurs.

Homologous series   A family of organic compounds which have similar structure, name endings and chemical properties. They show a trend in physical properties and their formula can be represented by a general formula, for example, alkanes C_nH_{2n+2}.

Hydrocarbon   A molecule which contains atoms of carbon and hydrogen only.

Hydroelectric power   Electricity generated by falling water turning turbines.

Hypothesis   A suggested explanation for a group of facts or phenomena.

Igneous rocks   Rocks formed when hot magma from the Earth's mantle cools and hardens. They are usually crystalline and are of two types, intrusive and extrusive. Intrusive igneous rocks, e.g. granite, are formed by crystallization of the magma underground. Extrusive igneous rocks, e.g. basalt, are formed by crystallization of the magma on the Earth's surface.

Impact strength   A measure of the ability of a substance to withstand an impact.

Incomplete combustion   The incomplete oxidation of a hydrocarbon resulting in the formation of carbon monoxide and water.

Indicator   A substance used to show whether a substance is acidic or alkaline (basic) e.g. phenolphthalein.

Inert gases   Unreactive gaseous elements found in group 0 of the periodic table.

Inner core   Solid rock at the centre of the Earth at very high temperature and pressure, composed of mainly nickel and iron. It has a diameter of 2530km.

Insoluble   If the solute does not dissolve in the solvent it is said to be insoluble.

Intermolecular bonds/forces (van der Waals' forces)   Weak attractive forces which act between molecules, for example van der Waals' forces.

Intramolecular bonds   Forces which act within a molecule e.g. covalent bonds.

Intrusive igneous rocks   Igneous rocks formed by molten rock being forced up into the Earth's crust but not breaking the surface.

Ion   An atom or group of atoms which has either lost one or more electrons making it positively charged or gained one or more electrons making it negatively charged.

Ion exchange   The process in which one ion is exchanged for another ion.

Ionic bonds   A strong electrostatic force of attraction between oppositely charged ions.

Ionic equation   The simplified equation which we can write if the chemicals involved are ionic.

Irritant   A substance which causes irritation.

Isotopes   Atoms of the same element which possess different numbers of neutrons. They differ in mass number (nucleon number).

Key   A method of sorting.

Kinetic theory   A theory which accounts for the bulk properties of matter in terms of the constituent particles.

Latex   A white sticky liquid collected from rubber trees which consists of long polymer chains.

Lava   Magma which has reached the surface of the Earth.

Law of conservation of mass   For any chemical reaction the total mass of reactants is equal to the total mass of products produced.

Lime   A white solid known chemically as calcium oxide (CaO). It is produced by heating limestone. It is used to cure soil acidity and to manufacture calcium hydroxide (slaked lime). It is also used as a drying agent in industry.

Limestone   A form of calcium carbonate ($CaCO_3$). Other forms include chalk, calcite and marble. It is used directly to neutralise soil acidity, in the manufacture of iron and steel, glass, cement, concrete, sodium carbonate, and lime.

Liquid   A state of matter intermediate between a solid and a gas.

Liquified petroleum gas (LPG)   A mixture of the hydrocarbons propane and butane used as a domestic and industrial fuel.

Lubricant   Any substance used to reduce friction between two moving surfaces.

Macromolecules   Very large molecules.

Magma   Is the molten rock material containing dissolved gases as well as water beneath the Earth's crust.

Magnetic   Can be attracted to a magnet.

Malleable   The property of a metal which allows it to be hammered into thin sheets without breaking.

Mantle   A thick layer of solid, dense rock rich in magnesium and silicon which surrounds the outer core of the Earth.

Manufactured   A substance which has been produced from raw materials, for example, copper is extracted from its ore malachite.

Mass   The amount of matter in a substance, measured in kilograms.

Mass number (nucleon number)   The total number of protons and neutrons found in the nucleus of an atom, symbol A.

Matter   Anything which occupies space and has a mass.

Melting point   The temperature at which a solid begins to liquefy. Pure substances have a sharp melting point.

Membrane cell   An electrolytic cell used for the production of sodium hydroxide, hydrogen and chlorine from brine in which the anode and cathode are separated by a membrane.

Metallic bond   An electrostatic force of attraction between the mobile 'sea' of electrons and the regular array of positive metal ions within a solid metal.

Metalloids (semi-metal)   Any of the class of chemical elements intermediate in properties between metals and non-metals, for example, boron and silicon.

Metals   A class of chemical elements which have a characteristic lustrous appearance and which are good conductors of heat and electricity.

Metamorphic rocks   These are rocks formed when rocks buried deep beneath the Earth's surface are altered by the action of great heat and pressure. For example, marble is a metamorphic rock and is formed from limestone by this type of action.

Microbes   Any microscopic organism.

Micro-organism   Any organism such as a bacterium, virus or protozoan of microscopic size.

Mineral   A naturally occurring substance of which rocks are made.

Molar mass   The mass of one mole of a compound.

Mole (mol) 6×10^{23} atoms, ions or molecules. This number is called Avogadro's constant.

Molecules   Groups of atoms chemically bonded together.

Monatomic molecules   A molecule which consists of only one atom e.g. neon, argon.

Monomer   A simple molecule, such as ethene, which can be polymerised.

Negatively charged   Having a negative electrical charge.

Neutral   A substance with a pH of 7, neither acidic nor alkaline.

Neutral (uncharged)   Having no electrical charge.

Neutralise   The process in which the acidity of a substance is destroyed. Destroying acidity means removing $H^+(aq)$ ions by reaction with a base, carbonate or metal.

Neutron   A fundamental, uncharged sub-atomic particle present in the nuclei of atoms.

Nitrogen cycle   The system by which nitrogen and its compounds both in the air and soil are interchanged.

Nitrogen-fixing bacteria   Bacteria present in root nodules of certain plants which are able to take nitrogen directly from the atmosphere to form essential protein molecules.

Nitrogenous fertiliser   A substance which is added to the soil to increase the amount of the element nitrogen.

Non-metals   A class of chemical elements that are typically poor conductors of heat and electricity.

Nucleus   The dense central part of an atom which contains the protons and neutrons.

Ocean-floor spreading   The process which increases the width of an ocean by magma emerging and pushing older rock apart.

Oceanic crust   The Earth's crust under the oceans.

Ore   A naturally occurring mineral from which a metal can be extracted.

Organic compounds   Compounds which contain the element carbon often bonded to hydrogen and also to other elements such as oxygen, nitrogen, the halogens, sulphur and phosphorus.

Outer core   Very dense liquid rock at high temperature composed of nickel and iron. The Earth's magnetic field arises here. It extends to a diameter of 6930 km.

Oxidation   The addition of oxygen to, or the loss of electrons from, a substance.

Oxidising agent   A substance which brings about oxidation.

Particles   Extremely small pieces of matter, for example, atoms, molecules and ions.

Period   Units of geological time during which systems of rocks are formed.

Period   Horizontal row of the periodic table. Within a period the atoms of all the elements have the same number of occupied energy levels but have an increasing number of electrons in the outer energy level.

Periodic table   A table of elements arranged in order of increasing atomic number to show the similarities of the chemical elements with related electron structures.

pH scale   A scale running from 0–14 used for expressing the acidity or alkalinity of a solution.

Photosynthesis   The chemical process by which green plants synthesise their carbon from atmospheric carbon dioxide using light as the energy source and chlorophyll as the catalyst.

Planet   An object which revolves around a star and which is illuminated by that star.

Plastics   Most plastics are polymers and are classified as either thermosetting or thermosoftening materials.

Polymer   A substance possessing very large molecules consisting of repeated units or monomers. Polymers, therefore, have a very large relative molecular mass.

Polymerisation   The chemical reaction in which molecules (the monomers) join together to form a polymer.

Polythene   A polymer formed from the monomer ethene.

Positively charged   Having a positive electrical charge.

Precipitate   An insoluble compound formed from solution during a chemical reaction.

Primary atmosphere   The original thick layer of gases, mainly hydrogen and helium, that surrounded the core soon after the planet was formed 4500 million years ago.

Products   Substances produced in a chemical reaction.

Protein   A substance formed from the reaction between many amino acids. For a molecule to be a protein there must be at least 100 amino acids involved.

Protons   A fundamental sub-atomic particle which has a positive charge, equal in magnitude to that of an electron. The proton occurs in all nuclei.

Quarrying   Open surface excavation for the extraction of minerals or ores.

Radiation   The way in which energy in the form of electromagnetic waves (mostly infra-red) flows from one place to another.

Radioactive   Atoms which have unstable nuclei which disintegrate spontaneously to give off one or more types of radiation are said to be radioactive.

Radioactive isotopes   Atoms which have unstable nuclei which disintegrate spontaneously to give off one or more types of radiation.

Random   Haphazard or without order.

Rate of reaction   The rate at which a reactant is used up or a product is formed in unit time.

Raw material   The starting material from which a more refined substance is produced.

Reactants   The starting materials in a chemical reaction.

Reactivity   The property of a chemical which determines how readily it takes part in a chemical reaction.

Reactivity series   An order of reactivity, giving the most reactive metal first, based on results from experiments with oxygen, water and dilute hydrochloric acid.

Redox   A term applied to any chemical process which involves both reduction and oxidation.

Reducing agent   A substance which brings about reduction.

Reduction   The removal of oxygen from, or the addition of electrons to, a substance.

Relative atomic mass   Symbol A_r.

$$A_r = \frac{\text{Average mass of isotopes of the element}}{1/12 \times \text{mass of a carbon-12 atom}}$$

Hence this gives rise to the relative atomic mass scale.

Relative formula mass (relative molecular mass)   This is the sum of the relative atomic masses of all those elements shown in the formula of the substance. This is often referred to as the relative molecular mass (M_r).

Respiration   It is a chemical reaction in which glucose is broken down in a cell and energy is released. Respiration takes place in all living organisms at all times.

Reversible reaction   A chemical reaction which is said to be reversible can go both ways. This means that once some of the products have been formed they will undergo a chemical change once more to reform the reactants. The reaction from left to right, as the equation for the reaction is written is known as the forward reaction and the reaction from right to left is known as the back reaction.

Rift valley   A long narrow valley formed from the subsidence of land between two parallel faults.

Rock cycle   The cycle of natural rock change in which rocks are lifted, eroded, transported, deposited and possibly changed into another type of rock and then uplifted to start a new cycle.

Rusting   Rust is a loose, orange/brown, flaky layer of hydrated iron(III) oxide found on the surface of iron or steel. The conditions necessary for rusting to take place are the presence of oxygen and water. The rusting process is encouraged by other substances such as salt. It is an oxidation process.

Salts   Substances formed when the hydrogen of an acid is completely replaced by a metal or the ammonium ion (NH_4^+).

Saturated compounds   Molecules which possess only single covalent bonds.

Saturated solution   This is a solution which contains as much dissolved solute as it can at a particular temperature.

Secondary atmosphere   Early volcanic activity created this mixture of gases. The mixture which formed this atmosphere included ammonia, nitrogen, methane, carbon monoxide, carbon dioxide and sulphur dioxide gases.

Sediment   Matter that settles to the bottom of rivers, seas or oceans.

Sedimentary rock   Rock which is formed when solid particles carried or transported in seas or rivers are deposited. Layers of sediment pile up over millions of years and the pressure created on the sediments at the bottom causes the grains to be cemented together. For example, limestone is a sedimentary rock.

Silt   Fine deposit of small particles of rock.

Simple molecular structures These substances possess between two and one hundred atoms per molecule.

Single covalent bond A chemical bond formed by the sharing of one pair of electrons between two non-metal atoms.

Slaking The addition of water to lime.

Solar activity High energy radiation created on the surface of the Sun which affects the atmospheres of neighbouring planets.

Solid One of the three states of matter. In a solid the particles are arranged in a regular manner and they are only able to vibrate about a fixed position.

Soluble If the solute dissolves in the solvent it is said to be soluble.

Solute The substance which dissolves in a solvent.

Solution This is formed when a substance (solute) disappears into (dissolves) another substance (solvent).

Solvent The substance in which the solute dissolves.

Spoilage The decay of food by the action of chemicals or micro-organisms.

States of matter Solid, liquid or gas.

Sub-atomic particles Those particles found within an atom.

Sublimation The direct change of state from solid to gas and the reverse process.

Tectonic plates The sections of the Earth's crust which move slowly about the surface of the Earth. The driving force behind the movement is thought to be convection currents in the mantle.

Tensile strength A measure of the ability of a material to withstand a stretching force.

Thermal decomposition The chemical breakdown of a substance under the influence of heat.

Thermoplastics These are plastics which soften when heated (e.g. polyethene, PVC).

Thermosetting plastics These are plastics which do not soften on heating but only char and decompose (e.g. melamine).

Titration A method of volumetric analysis in which a volume of one reagent (usually an acid) is added to a known volume of another reagent (usually an alkali) slowly from a burette until an end-point is reached. If an acid and alkali are used then an indicator is used to show that the end-point has been reached.

Transition elements A block of metallic elements situated between groups 2 and 3 in the periodic table.

Ultra violet radiation A high energy electromagnetic radiation just beyond the violet region of the visible spectrum.

Universal indicator A mixture of many other indicators. The colours shown by this indicator can be matched against the pH scale.

Unsaturated compound A compound which contains one or more double covalent bonds.

Valency The combining power of an atom or group of atoms. The valency of an ion is equal to its charge.

Vents The openings in the volcano through which volcanic gases, water vapour, carbon dioxide etc., rise to the surface.

Vulcanising This is the process which makes rubber harder and increases its elasticity. This is done by the formation of sulphur bridges between the rubber chains.

Weathering The action of wind, rain and frost on rock which leads to its erosion.

X-ray diffraction A technique used to study crystal structure.

X-rays A very high energy electromagnetic radiation.

Yeast Unicellular micro-organism capable of converting sugar into alcohol and carbon dioxide.

Yield The quantity of a product obtained in a chemical reaction.

Data tables

Indicators

	Acid	Alkaline
Thymol blue	red	yellow
Methyl yellow	red	yellow
Methyl orange	red	yellow
Bromophenol blue	yellow	blue
Congo red	violet	red
Bromocresol green	yellow	blue
Bromothymol blue	yellow	blue
Phenol red	yellow	red
Phenolphthalein	colourless	red
Universal* (solution)	red	purple

*Universal indicator is a mixture of several indicators.

Energy needed to break bonds

Bond	Relative amount of energy needed (kJ mol^{-1})
C — C	347
C = C	612
C — H	413
C — Cl	346
C — O	336
C = O	805
C — N	286
N — N	158
N = N	410
N ≡ N	945
H — H	436
H — N	391
H — O	464
H — Cl	432
H — Br	366
H — I	298
Cl — Cl	243
Br — Br	193
I — I	151

Colour changes for universal indicator

Colour	Red	Orange	Yellow	Green	Blue	Navy blue	Violet
pH	0–2	3–4	5–6	7	8–9	10–12	13–14

ACID	NEUTRAL	ALKALINE

Solubility of salts and hydroxides in water at room temperature

Soluble	Insoluble
All sodium, potassium and ammonium salts	
All nitrates	
Most chlorides, bromides, iodides	Silver and lead chlorides, bromides, iodides
Most sulphates	Lead sulphate, barium sulphate, calcium sulphate*
Sodium, potassium and ammonium carbonates	Most other carbonates
Sodium, potassium and ammonium hydroxides.	Most other hydroxides
Calcium hydroxide is slightly soluble	

*Calcium sulphate is slightly soluble and is present in some natural waters.

Names and formulae of some common ions

Positive ions Name	Formula	Negative ions Name	Formula
Hydrogen	H^+	Chloride	Cl^-
Sodium	Na^+	Bromide	Br^-
Silver	Ag^+	Fluoride	F^-
Potassium	K^+	Iodide	I^-
Lithium	Li^+	Hydroxide	OH^-
Ammonium	NH_4^+	Nitrate	NO_3^-
Barium	Ba^{2+}	Oxide	O^{2-}
Calcium	Ca^{2+}	Sulphide	S^{2-}
Copper(II)	Cu^{2+}	Sulphate	SO_4^{2-}
Magnesium	Mg^{2+}	Carbonate	CO_3^{2-}
Zinc	Zn^{2+}	Hydrogen-carbonate	HCO_3^-
Lead	Pb^{2+}		
Iron(II)	Fe^{2+}		
Iron(III)	Fe^{3+}		
Aluminium	Al^{3+}		

Properties of some common substances (under normal conditions)

Name	Melting point/°C	Boiling point/°C	Electrical conductivity (molten)	Density/ g cm^{-3} (x1000 kg m^{-3})	Structure
Aluminium oxide	2072	2980	good	3.97	giant
Ammonia	−77	−34	poor	0.0007	molecular
Barium chloride	963	1560	good	3.9	giant
Calcium chloride	782	1600	good	2.15	giant
Calcium oxide	2614	2850	good	3.35	giant
Carbon dioxide	sublimes	−78	poor	0.0018	molecular
Carbon monoxide	−199	−191	poor	0.0012	molecular
Copper(II) chloride	620	993	good	3.39	giant
Copper(II) sulphate	200	decomposes	—	2.28	giant
Hydrogen chloride	−114	−85	poor	0.0015	molecular
Iron(III) oxide	1565	—	good	5.2	giant
Lead chloride	500	950	good	5.85	giant
Lithium chloride	605	1340	good	2.07	giant
Lubricating oil	—	250–350	poor	0.8*	molecular
Magnesium chloride	714	1412	good	2.32	giant
Methane (natural gas)	−182	−161	poor	0.00066	molecular
Methylated spirit	−100	80	poor	0.79	molecular
Paraffin wax	50-60	decomposes	poor	0.8*	molecular
Petrol	−60 to −40	40 to 75	poor	0.6	molecular
Potassium chloride	770	1500*	good	1.98	giant
Silicon dioxide (sand)	1610	2230	poor	2.65	giant
Sodium chloride	801	1413	good	2.17	giant
Sugar (sucrose)	161	decomposes	poor	1.5	molecular
Sulphur dioxide	−73	−10	poor	0.0026	molecular
Water	0	100	poor	1.00	molecular

*Approximate values.

Properties of elements

Element	Melting point/°C	Boiling point/°C	Density/ g cm^{-3} (x1000 kg m^{-3})
Aluminium	660	2470	2.7
Argon	−189	−186	0.0016
Barium	725	1640	3.5
Bromine	−7	59	3.1
Caesium	29	669	1.9
Calcium	840	1484	1.5
Carbon			
*Graphite	sublimes	4800	2.3
*Diamond	sublimes	4800	3.5
Chlorine	−101	−35	0.0029
Copper	1084	2570	8.9
Fluorine	−220	−188	0.0016
Germanium	937	2830	5.4
Gold	1064	3080	19.3
Helium	−272	−269	0.00017
Hydrogen	−259	−253	0.00008
Iodine	114	184	4.9
Iron	1540	2750	7.9
Lead	327	1740	11.3
Lithium	180	1340	0.5
Magnesium	650	1110	1.7
Mercury	−39	357	13.6
Nitrogen	−210	−196	0.0012
Oxygen	−218	−183	0.0013
Phosphorous (white)	44	280	1.8
Potassium	63	760	0.86
Rubidium	39	686	1.5
Silicon	1410	2355	2.32
Silver	960	2212	10.5
Sodium	98	880	0.97
Strontium	769	1384	2.6
Sulphur (rhombic)	113	445	2.1
Tin (grey)	232	2270	7.3
Titanium	1660	3290	4.5
Zinc	420	907	7.1

*Carbon (as either graphite or diamond) does not melt but sublimes (i.e. changes directly from solid to gas); the temperature is approximate. Gas densities are quoted at 25°C and atmospheric pressure.

Common ores of metals

Name of ore	Chemical formula	Metal extracted
Haematite	Fe_2O_3	Iron
Bauxite	Al_2O_3	Aluminium
Halite	$NaCl$	Sodium
Zinc blende	ZnS	Zinc
Malachite	$Cu_2CO_3(OH)_2$	Copper
Copper pyrites (chalcopyrite)	$CuFeS_2$	Copper
Ilmenite	$FeTiO_3$	Titanium
Wolframite	$FeWO_4$	Tungsten
Cassiterite	SnO_2	Tin
Galena	PbS	Lead

Tests for common gases

1 Hydrogen
   If a test tube of hydrogen is held to a flame, the hydrogen burns in air with a squeaky 'pop'.
2 Oxygen
   If the glowing tip of a wooden spill is put into a test tube of oxygen, the spill will burst into flame.
3 Carbon dioxide
   If carbon dioxide is bubbled through limewater (calcium hydroxide solution), a white solid (precipitate) is formed and the solution appears cloudy or milky.
4 Chlorine*
   Chlorine is a yellow-green gas which can be recognised by its smell. It bleaches damp litmus paper and turns damp starch-iodide paper blue-black.
5 Sulphur dioxide*
   It changes the colour of filter paper dipped into an acidified potassium dichromate solution from orange to green.
6 Ammonia*
   Ammonia can be recognised by its smell; if the gas is tested with a damp indicator paper, the indicator paper turns to the alkaline colour (blue with universal indicator or with litmus).

*These gases are *poisonous* so be very careful.

Tests for ions

Ion	Test
Chloride (Cl^-)	Add a few drops of nitric acid and then a few drops of silver nitrate solution. A white precipitate is formed.
Sulphate (SO_4^{2-})	Add a few drops of nitric acid and then a few drops of barium chloride solution. A white precipitate is formed.
Carbonate (CO_3^{2-})	Add dilute hydrochloric acid to the solid (or mix with the solution). Bubbles of gas are given off.
Iron(II) (Fe^{2+})	Add aqueous sodium hydroxide to the solution. A pale green jelly-like precipitate is formed.
Iron(III) (Fe^{3+})	Add aqueous sodium hydroxide to the solution. A brown jelly-like precipitate is formed.
Copper(II) (Cu^{2+})	Add aqueous ammonia to the solution of copper(II) ions. A pale blue precipitate is formed; this dissolves when more ammonia solution is added and a deep blue solution is formed.

Fractions from crude oil and their uses

Name of fraction	State (under normal conditions)	Boiling point range (approximate)	Uses	Length of carbon chain (approximate)
Petroleum gases	gas	up to 25°C	Calor gas; Camping Gaz	$C_1 - C_4$
Petrol	most runny liquid	40–100°C	Petrol for cars	$C_4 - C_{12}$
Naphtha		90–150°C	Petrochemical	$C_7 - C_{14}$
Kerosene		150–240°C	Jet fuel; petrochemicals	$C_9 - C_{16}$
Diesel		220–300°C	Central heating fuel; petrochemicals	$C_{15} - C_{25}$
Lubricating oil	least runny liquid	over 350°C	Lubricating oils; petrochemicals	$C_{20} - C_{70}$
Bitumen	solid	over 400°C	Roofing; surfacing roads	C_{50} and up

Index

A$_r$, see relative atomic mass
acid rain 32,82–3,146
acidity,pH measurement of 33
acids
    properties 32
    reactions 13,35,36,37–9
    strong 32
    weak 32
activation energy 159,162,163
addition polymerisation 89
aerobic respiration 154
air, see atmosphere
alcohols 155
algae 62
alkali metals 129,132–3
alkaline earth metals 129
alkalinity,pH measurement of 33
alkalis
    properties 32
    reactions 35
    strong 32
    weak 32
alkanes 84–6
alkenes 87
allotropes 120
alloy 6,125
aluminium
    alloy 6
    environmental cost 29
    extraction 19,28–9
    properties 15,28
    reactions 16,37
ammonia
    manufacture 168–9
    molecule 118
    solution 38
    uses 164,169
ammonium
    chloride 36
    compounds 164,166–7
    nitrate 166–7,169
    sulphate 36
anaerobic respiration 154
anion 113
anode 26,127
argon 111
artificial fertilisers 164–5,167,168
atmosphere
    composition 80
    formation 62–3
    pollution 80–1
    primary 62
    secondary 62
atomic mass, relative 173
atomic mass unit 108
atomic number 108
atomic theory 106,110
atoms 2, 106
    and kinetic theory 100
    structure of 108–9,110–111
Avogadro's constant 173
Avogadro's Law 176

bacteria 62
baking 155
balanced chemical equations 10, 180
base 38–9
basic oxygen furnace 21
bauxite 28
blast furnace 20
bleaching agent 152
boiling 103
boiling point
    and distillation 76–7
    examples listed
        3,86,87,96,97,133,171
bonding
    chemical 113
    covalent 86,106,118–9
    ionic 113–15,116
    metallic 124
bonds 106
    breaking and formation 161
    chemical 161
    energy of 161
    giant molecular 120
    inter/intramolecular 120
    simple molecular 120
    van der Waals' 86,120
brewing 154
bromine 134
butane 84,86
butene 87

caesium chloride 116,117
calcium 15
calcium carbonate (limestone)
    formation 65–6
    reactions 20,146,148
    uses 36,46,47,148
calcium chloride 36,114–115
calcium hydroxide (slaked lime)
    manufacture 47–8
    solution (limewater) 38
    uses 48
calcium oxide (lime,quicklime)
    manufacture 47,180
    reactions 20,21,48,
calcium sulphate 36,46
canyons 52
carbon
    allotropes 120
    alloys, see also steel, 7
    properties 18–19
    source 45
    structures 112
carbon dioxide
    as pollutant 82
    in air 45,62–3,78,80
    reactions 20
carbon monoxide 20,152
carbonates
    fermentation 36
    reaction with acids 37
carbonic acid 32,36,82

catalysts
    defined 85,152,163,171
    enzymes as 152,154,157,163
    examples 85,137,168
catalytic converter 152
cathode 26,127
cation 114
cement manufacture 46
charge, electric 107
chemical bonds
    defined as 113
    making and breaking 161
chemical change 8,9
chemical energy 160–163
chemical equations 10
chemical equilibrium 170
chemical formula 10
chemical reactions 8–11
chloride ion 135
chlorides 36
chlorine
    manufacture 126–7
    properties 131,134
    reaction 134
    uses
        water treatment 134
chlorophyll 150
chlor-alkali industry 126–7
chromium 7,16,137
clock reactions 145
coal 63,74
combustion
    defined 78,158
    energy release 160,161
    heat of 162
    products 79,158
competition reaction 19
complete combustion 160
compounds
    covalent 118–123
    defined 3,107
    ionic 116
compression of liquids and gases 99
concentration of solutions 176–177
    and reaction speed 146,149
concrete 47,114–15
condensation 77,97
conduction electrons 124
conduction ions 27
continental crust 55
continental drift, theory of 55–7
contraction 98
convection currents 56
copper
    extraction 30–31
    ion 117
    properties 15,30,137
    reactivity 17
copper carbonate 9
copper(II) oxide 9,39
copper(II) sulphate 31
core, Earth's 54

corrosion 22
covalent bond 86,87,118–119
covalent compounds 118–123
cracking 75,84–5
cross linking 121
crude oil 74,76,84–5
crust, Earth's
   composition 54
   raw materials from 44,72
crystalline 113
crystallization 37
crystal lattice 116,125

decay
   microbes 156
   prevention 157
delta 64
density 72
deposition of sediments 64
diamond
   molecular structure 120
   properties 112
diatomic molecules 11,106,134
diesel 75–7
diffusion, gaseous 95,100–101
displacement reactions 17,134
dissolving 104,162
distillation, separation method 76–7
double covalent bond 87
drug 143
ductility 4,125
dynamic equilibrium 170

Earth's crust 9
Earth's structure 54
earthquake 58
effervescence 13
electric charge 107
electrical conduction 116,124
electrodes 26
electrolysis
   defined 19,26
   of aluminium oxide 28–9
   of sodium chloride solution 126–7
electrolytes 26,126
electron 'sea' 124
electrons
   arrangement in atoms 108,110
   configuration 110
   shells 108,110–111
   structure 110,130–1
   transferring of 25
electroplating
   process 26
   and rust prevention 24
electrostatic force 108
elements
   defined 2,106
   examples 3,109,111
   symbols 128
empirical formula 178
end-point 38
endothermic reaction 159,162
energy
   bond 161
   and change of state 102–103

level diagrams 160,162,163
levels 108,110–111,118
enthalpy
   change 162
   of combustion 163
enzymes 142,152,154,157,163
epoch 67
equilibrium, chemical 170
era 66–7
erosion of rocks 52
ethane 75,86
ethanoic acid 32,33,36,119
ethanol 85,154
ethene 85,87
evaporation 37,38,97,102–103
exothermic reaction 16,158–162
expansion 98
extraction
   of aluminium 19,28–9
   of iron 20
   of metals 18–21
extrusive rocks 60

faults 51
fermentation 154–5
fertilisers
   artificial 164,167
   pollutants 165
   production 167,169
fibres 121
filtration 37
fissures 60
fluorine 131,137
fold mountains 57
food
   preservation 157
   spoilage 156
formulae
   compounds 117
   empirical 178
   ionic 117
   molecular 179
fossil fuels 45,63,74–5,78
fossils 50,56,65–7
fractional distillation 76–7
fractions 76,78
freezing 96
friction 112
fuels 78–9,84,160

galvanising 25
gas (natural) 74
gas(es) (general) 94–5
geological time scale 66–7
   and periods 66
giant ionic structures 116
giant molecules 120
glaciers 51
glass 47
global warming 81
glucose 104,107
gold 12,15
graphite 112,120
greasing 24
greenhouse effect 81
groups of elements 129–137

Haber process 168,171
haematite 18,20
Hall-Héroult cell 29
Halogens
   and periodic table 129
   properties 134
   reactivity 135
heat of reaction 162
helium 111,136
homologous series 86
hydrocarbons 74–7,79
hydrochloric acid
   properties 32,33
   reactions 35,73,101,148
hydroelectricity 29
hydrogen
   covalent bonds 118
   manufacture 126–7
   properties 137
   uses 168
hydrogen peroxide 153
hydrogencarbonate 37

ice-pack 159
igneous rocks 60–1,71
incomplete combustion 160
indicators 32–3,38
inert gases 111,129
inter/intramolecular bonds 120
intermolecular forces 86
insoluble 105
intrusive rocks 61
iodine 134
   structure 120
ion exchange 126
ionic
   bonding 113–115
   compounds 116
   equations 35,38
   lattice 35,114,116
   structure 114,116
ions 100,107,114
   and electrolysis 26
iron
   alloys, see also steel,
   as catalyst 170–1
   extraction 20–21
   ion 117
   properties 12,14,15,137
   reactions 16,37
   reactivity 22
iron(II) oxide 16,20
iron(II) sulphate 36
isotopes 109,173

Kerosene 75–77
key, identifying rocks 71,72
kinetic theory 100–101

lactic acid 156
latex 122
lattice 116,125
lava 60
law of conservation of mass 180
lead 15
lead bromide 26

light, and reactions 150
lime (quicklime), *see* calcium oxide
lime (slaked), *see* calcium hydroxide
limestone, *see* calcium carbonate
limewater, *see* calcium hydroxide solution
liquefied petroleum gas 84
liquid 94–5
lithium
   and periodic table 130
   properties 132–3,137
lubrication 112

macromolecule 120
magma 56,58,59,60,68,70
magnesium
   alloy 6
   reactions 11,13,37,178
   properties 14,15,25
magnesium oxide 13
magnetic stripes 59
magnetite 20
malachite 9,44
malleability 4,125
manganese alloy 7
manganese(IV) oxide 153
mantle, Earth's 54
mass
   law of conservation of 180
   molar 174
   number 108
mass spectrometer 179
materials 44–5
'MAZIT' metals 37
melting 102
melting point 96,97
   examples listed 3,86,87,133,171
membrane cell 126
Mendeléev, Dmitri 128
mercury 19
metallic bond 124
metalloids 129
metals
   extraction from ores 18–21
   properties 3,4,6,124–5
   reactions 12–15,36–7
   reactivity series 14,15
   structure 124–5
   uses 6,7
metamorphic rocks 58,68–9,70,71
methane
   as fuel 79,84,158
   molecule 86,118–119
   pollutant 81
microbes 156–7
micro-organisms 156
minerals 9,44,72
mixture 104
molar heat of combustion 163
molar mass 174
mole
   and chemical equations 180
   and compounds 174
   and elements 173
   and gases 176
   and solutions 176
   defined 173

molecular
   formula 179
   structure 112
molecules 100,106–7
monatomic molecules 136
monomer 89,121

negative (photographic) 150
neon 111,136
neutralisation of acid 34–39,167
neutrons 108
nitrates
   fertilisers 165
   formation 36
nitric acid 32,36,82
   production 168–9
   reaction 37,167
nitrogen
   as pollutant 82
   in air 80,166
   use
      fertiliser 165
      industrial 168
nitrogen cycle 166
nitrogen dioxide 169
nitrogen-fixing bacteria 166
nitrogen monoxide 152,162,169
nitrogenous fertilisers 167
noble gases 111,129,136
non-metals 3,4
nucleon number 108–109
nucleus 108,110

ocean resources 45
oceanic crust 55
ocean-floor spreading 58–9
oil
   as fuel 74
   formation 74
oiling 24
ores 9,18,44
organic compounds 75,178
oxidation reactions 79,127,156
   and reduction 127
   defined 16
oxide layer 15
oxides 13
oxidising agent 16
oxygen
   in air 45,61,80,106
   reactions 11,13,132
ozone layer 62–3

paint 123
painting 24
particles 95,97,98,100,106
   and rate of reaction 146–150
periodic table
   history 128
   periods 129
   position of elements 128
pH scale 33
phosphorous
   in fertiliser 165
photographs 150
photosynthesis 62,80,150,159,162

physical state symbols 10
plastics 121
plate tectonics 58–9
plating 24
platinum 15
plutonium 3
pollution in air 80–81
polyethene (polythene) 85,88,89
polymer 88,121
polymerisation 89
polypropene 89
potassium
   and periodic table 130
   properties 12,15,132–3
   use
      in fertiliser 164–5
potassium hydroxide 32,38
potassium nitrate 36
precipitation of salts 149
pressure
   and speed of reaction 144
primary atmosphere 62
products 9,144,180
propane 84,86
propene 87
protein 154
proton 108
proton number 108,109,111,128–9
purification of copper 30–31

quarrying 48
quicklime, *see* calcium oxide

radiation, Earth's 54
radioactive elements 2,129
radioactive isotopes 54,109
   Earth's 54
rain
   acid 32,83–4
random motion 100
rates of reaction 142–7
   catalysts 144,152
   concentration 144,146,149
   light 144,150
   pressure 144
   surface area 144,148
   temperature 144,150,159
   yield 170
raw materials 3,44,72,74
reactants 9,144,180
reaction, heat of 162
reactivity 16
   series 14,15,18
   trends 133,135
redox reactions 16,25
reducing agent 16
reduction 16
   of oxides 18
   with carbon 19
relative atomic mass 173
relative formula mass 174
relative molecular mass 174
respiration 63,80,118,154,158
reversible reaction 169
rift valleys 51
rock cycle 70

rocks
    erosion 52
    weathering 52,65,68,70,82
rubber 122
rusting of iron 8,12,22–25
    prevention 24

sacrificial protection 25
salt
    common, see sodium chloride
    formation 126
    normal 35,36–37,38–39
saturated molecules 86
saturated solution 37
sedimentary rock 57,58,64–66,70–71
sediments 64
silica (silicon(IV) oxide, sand)
    structure 120
    uses 21
silt 64
silver
    properties 15
    use
        in photography 150
silver bromide, use 36
simple molecular structure 120
slag 21
slaked lime 48,83
sodium
    and periodic table 130
    properties 15,113,132–3
sodium carbonate 36,47
sodium chloride 35,107,113–114,117,
        126
sodium hydroxide 32,35,38
    manufacture 126–7
sodium stearate 36

sodium thiosulphate 149
solar activity 62
solid, properties 94–5
soluble 105
solute 104
solution 104
solvent 104
spectator ions 35
spoilage 156
states of matter 94,95,102
steel
    alloy 7
    manufacture 21
    rusting 22–25
structure of materials 112
sub-atomic particles 2
sublimation 96,97
sugar 118,154–5,156
sulphates, formation 36
sulphur
    and vulcanising 122
    as pollutant 82
sulphur dioxide 32,82–3,
sulphuric acid 32,33,36,82
    reactions 37,39,146
sulphurous acid 82
suspensions, liquid 105
symbols
    examples listed 3,109,111
    physical state 10

tectonic plates 55,58
temperature
    and expansion 98
    and kinetic theory 100
    and rate of reaction 150,159
test for unsaturation 87

thermal decomposition 10,20
thermit reaction 16
thermoplastics 121
thermosetting plastics 121
titration 38
transition metals
    and periodic table 129
    properties 137
trends (in periodic table) 133,134–5

universal indicator 33
unsaturated molecule 87

valency 117
van der Waals' forces 86,120
vents 60
volcanoes 50,60
vulcanising 122

water
    molecule 106,119
    properties 123
    reactions with metals 14,132
    states of matter 96
water treatment
    chlorination 134
weathering of rocks 52,65,68,70,82

X-ray diffraction 116

yeast 154–5
yield 169,170–1

zinc
    plating 25
    properties 14,15,25,137
    reactions 37

Periodic table

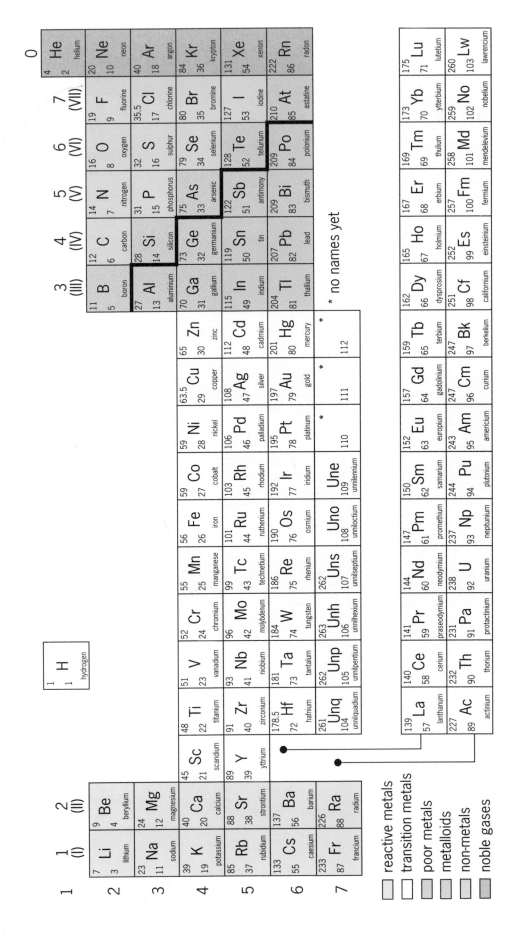

	1 (I)	2 (II)												3 (III)	4 (IV)	5 (V)	6 (VI)	7 (VII)	0
1					1 H hydrogen														4 He 2 helium
2	7 Li 3 lithium	9 Be 4 beryllium												11 B 5 boron	12 C 6 carbon	14 N 7 nitrogen	16 O 8 oxygen	19 F 9 fluorine	20 Ne 10 neon
3	23 Na 11 sodium	24 Mg 12 magnesium												27 Al 13 aluminium	28 Si 14 silicon	31 P 15 phosphorus	32 S 16 sulphur	35.5 Cl 17 chlorine	40 Ar 18 argon
4	39 K 19 potassium	40 Ca 20 calcium	45 Sc 21 scandium	48 Ti 22 titanium	51 V 23 vanadium	52 Cr 24 chromium	55 Mn 25 manganese	56 Fe 26 iron	59 Co 27 cobalt	59 Ni 28 nickel	63.5 Cu 29 copper	65 Zn 30 zinc		70 Ga 31 gallium	73 Ge 32 germanium	75 As 33 arsenic	79 Se 34 selenium	80 Br 35 bromine	84 Kr 36 krypton
5	85 Rb 37 rubidium	88 Sr 38 strontium	89 Y 39 yttrium	91 Zr 40 zirconium	93 Nb 41 niobium	96 Mo 42 molybdenum	99 Tc 43 technetium	101 Ru 44 ruthenium	103 Rh 45 rhodium	106 Pd 46 palladium	108 Ag 47 silver	112 Cd 48 cadmium		115 In 49 indium	119 Sn 50 tin	122 Sb 51 antimony	128 Te 52 tellurium	127 I 53 iodine	131 Xe 54 xenon
6	133 Cs 55 caesium	137 Ba 56 barium	139 La 57 lanthanum	178.5 Hf 72 hafnium	181 Ta 73 tantalum	184 W 74 tungsten	186 Re 75 rhenium	190 Os 76 osmium	192 Ir 77 iridium	195 Pt 78 platinum	197 Au 79 gold	201 Hg 80 mercury		204 Tl 81 thallium	207 Pb 82 lead	209 Bi 83 bismuth	209 Po 84 polonium	210 At 85 astatine	222 Rn 86 radon
7	233 Fr 87 francium	226 Ra 88 radium	227 Ac 89 actinium	261 Unq 104 unnilquadium	262 Unp 105 unnilpentium	263 Unh 106 unnilhexium	262 Uns 107 unnilseptium	Uno 108 unniloctium	Une 109 unnilennium	* 110	* 111	* 112							

* no names yet

140 Ce 58 cerium	141 Pr 59 praseodymium	144 Nd 60 neodymium	147 Pm 61 promethium	150 Sm 62 samarium	152 Eu 63 europium	157 Gd 64 gadolinium	159 Tb 65 terbium	162 Dy 66 dysprosium	165 Ho 67 holmium	167 Er 68 erbium	169 Tm 69 thulium	173 Yb 70 ytterbium	175 Lu 71 lutetium
232 Th 90 thorium	231 Pa 91 protactinium	238 U 92 uranium	237 Np 93 neptunium	244 Pu 94 plutonium	243 Am 95 americium	247 Cm 96 curium	247 Bk 97 berkelium	251 Cf 98 californium	252 Es 99 einsteinium	257 Fm 100 fermium	258 Md 101 mendelevium	259 No 102 nobelium	260 Lw 103 lawrencium

reactive metals
transition metals
poor metals
metalloids
non-metals
noble gases